The Regulation of DNA Replication and Transcription
The Role of Trigger Molecules in
Normal and Malignant Gene Expression

Experimental Biology and Medicine

Monographs on Interdisciplinary Topics

Vol. 8

Series Editor
A. Wolsky, New York, N.Y.

Co-Editors
D.J. Pizzarello, New York, N.Y.
G.V. Sherbet, Newcastle upon Tyne
J. Steiner, Jerusalem

KARGER

S. Karger · Basel · München · Paris · London · New York · Tokyo · Sydney

The Regulation of DNA Replication and Transcription

The Role of Trigger Molecules in Normal and Malignant Gene Expression

Mirko Beljanski

Faculté de Pharmacie, Université Paris-Sud, Paris, France

29 figures and 2 tables, 1983

S. Karger · Basel · München · Paris · London · New York · Tokyo · Sydney

Experimental Biology and Medicine
Monographs on Interdisciplinary Topics

National Library of Medicine, Cataloging in Publication
Beljanski, Mirko
The regulation of DNA replication and transcription: the role of trigger molecules in normal and malignant gene expression/
Mirko Beljanski. — Basel; New York: Karger, 1983.
(Experimental biology and medicine; v. 8)
1. DNA Replication 2. Gene Expression Regulation 3. Transcription, Genetic
I. Title II. Series
W1 EX479 v. 8 [QH 450 B431r]
ISBN 3-8055-3631-3

Drug Dosage

The authors and the publisher have exerted every effort to ensure that drug selection and dosage set forth in this text are in accord with current recommendations and practice at the time of publication. However, in view of ongoing research, changes in government regulations, and the constant flow of information relating to drug therapy and drug reactions, the reader is urged to check the package insert for each drug for any change in indications and dosage and for added warnings and precautions. This is particularly important when the recommended agent is a new and/or infrequently employed drug.

Printed in Switzerland by Thür AG Offsetdruck, Pratteln
ISBN 3-8055-3631-3

Contents

Preface

This monograph has its origin in discussions I had with several scientists and particularly with Professor *Alexander Wolsky* from New York University. It came out that the classical model for gene repression and derepression does not account for a number of basic questions in biology. It also appears that the commonly used term of 'gene activation' has not received a satisfactory definition on the molecular level. Moreover, the notion of 'gene activation' appears not to be exactly the same in the mind of biochemists, embryologists or general biologists. This prompted me to attempt to look for scientific data in which a common denominator may be an essential part for explaining the differential release of information from DNA in different biological systems (subcellular systems, cells, organs or complex multicellular organisms). The chosen experimental results have been obtained when one of the above systems was submitted to the effect of either one or several biological or chemically synthesized molecules. The final aim of my efforts was to propose a model for a molecular mechanism of differential gene activation and inactivation. For such a model one must take into account the basic process of DNA replication and transcription. The present volume is devoted to this problem and contains many data from the literature and a large amount of our results obtained for many years in this field.

I wish to express my deep thanks to Professor *A. Wolsky* for his constant interest and critical reading of the manuscript. Many useful comments or suggestions were received from Professors *M.C. Niu* and *H. Slavkin,* to whom I also express my thanks. I am greatly indebted to my wife and collaborator, *Monique Beljanski,* for her constant help and many suggestions throughout the preparation and typing of the manuscript, and to Dr. *L. Le Goff* who has very generously helped me for the final preparation of the manuscript and several illustrations.

Introduction

In spite of a considerable progress accomplished by biologists, biochemists and embryologists during the last 20 years in order to understand the cell differentiation and tissue development, the mechanism of gene activation remained essentially unclarified. The main difficulties arise from a tremendous degree of conformational complexity of the DNA double-helical structure and from the fact that it is very difficult to analyze with accuracy numerous biochemical pathways interfering in gene activation or inactivation in situ. To this it should be added that extrapolation of data collected by biochemists with the in vitro biological systems cannot be automatically applied to living organisms. Cell complexity and cell-to-cell interaction constitute a puzzle for scientists since these events are connected with gene activation or inactivation and regulation of DNA replication.

Variable activity of genes in a constant genome is taken as a basis for studying gene activation or inactivation through cell differentiation and proliferation. The expression of gene activity is usually followed by the appearance either of specific protein or messenger RNA, both in close relationship with DNA replication. These studies are not easy to perform due to the existence in eukaryotes of high molecular weight precursor RNAs from which emerges a variety of 'mosaic' messenger RNAs.

According to the classical model of gene regulation, proteins are considered as repressors of genes. However, no evidence was presented to demonstrate the specificity of a given non-histone protein in the repression of a single gene. Derepressors would act on 'promoter' regions as the only target on DNA where RNA polymerase initiates transcription. Other models suggested the participation of RNAs at the transcriptional level. However, a basic molecular mechanism for gene activation was not proposed. These models do not explain the apparent stability of certain synthetic pathways which appear or disappear during cell differentiation. How is the newly created state maintained through successive division of these cells requiring the regulation of DNA replication? In addition, the loss of the normal regulation process in malignant cells harboring neither mutation nor chromosomal deletion point to the existence of a basic molecular mechanism for reversible activation and inactivation of genes. In addition, a given state of differentiation requires the integrated activation of a number of non-contiguous genes.

In several biological systems it was possible to demonstrate that chemically unrelated substances (their concentrations are of importance) may induce either cell differentiation (or dedifferentiation, or transformation) or accelerate (or prevent) cell proliferation. No matter if these substances (hormones, RNA, RNA fragments, alkaloids, actinomycin D, phorbol esters, etc.) come from living cells or from other sources (chemical carcinogens, drugs, dimethylsulfoxide, etc.) provided that they interact directly or indirectly with particular segments of the DNA systems. Their effects may lead to a new and stable cell phenotype without inducing mutations. It was suggested by some authors that persistent action of particular molecules and their cumulative effect on normal cells may result in the appearance of cancerous cells. On the other hand, evidence was presented that plant and mammalian teratoma cells may differentiate into normal tissues under the appropriate biological conditions. These observations demonstrate that cytoplasmic molecules of normal cells may correct the expression of genes connected with the regulation of DNA replication in cancer cells.

Analysis and comparison of a very large number of experimental data obtained with different biological systems in the absence or presence of chemically unrelated substances point to the existence of a basic underlaying biochemical mechanism involved in reversible activation and inactivation of genes. Correlation can be demonstrated between the effects of numerous substances on in vitro DNA replication, release of specific information, translation and in situ effects on cell differentiation, dedifferentiation or maintenance of cell activity in nondividing cells. In all these events there seems to exist a common denominator which is the basis of our molecular model accounting for the regulation of DNA replication and the principle of differential release of genetic information.

1. Initiation and Control of DNA Replication

Introduction

The steady-state character of nondividing cells requires the maintenance of all necessary molecules in order to conserve specific biochemical and morphological characteristics. In dividing cells, genetic material has to be correctly and completely replicated. The enzyme which carries out this task is DNA-dependent DNA polymerase. This enzyme is responsible for the polymerization of deoxyribonucleotides and uses parental DNA as template. The biosynthesis of DNA includes initiation, elongation and termination of newly synthesized DNA chains. This is termed replication of DNA. DNA-dependent DNA polymerase, concentrated essentially in the nucleus, is in an environment where a great number of endogenous and exogenous molecules are interfering although they do permit, if necessary, the replication or transcription of the DNA. All these events have to be performed normally without altering genes and cell structure. The high molecular weight of eukaryotic DNA located in the chromosomes containing nucleosomes makes any examination of the control of DNA replication difficult, particularly during cell differentiation. The activity of DNA depends on the physicochemical structure of the double-stranded DNA. This process requires the opening and closing of DNA chains, events in which endogenous and exogenously introduced molecules interfere.

Attempts have been made in this monograph to bring data together permitting the construction of a model of the molecular mechanism for gene activation and its opposite, the mechanism of gene inactivation which, of course, involves the whole problem of temporary changes in the physiological state of the nondividing cells and their regulation.

Biosynthesis of DNA

Deoxyribonucleic acids (DNAs) may be defined as macromolecules in which phosphate groups bind together the deoxyribonucleotides by means of phosphodiester linkages (3′,5′-deoxyribose). The ratios of the four DNA bases, two purines (G and A) and two pyrimidines (C and T), are following the G=C and A=T rule, but the (G+C)/(A+T) ratio varies with the species. Although the base composition of total DNA from higher animals and plants varies within narrow limits, the genome contains various nucleotide segments which differ markedly from each other in mean base composition [*Woese*, 1967; *Shugalin* et al., 1970]. The DNAs of animals and plants, compared to those of bacteria or viruses, are generally richer in AT bases than in GC bases. The concept of complementarity of bases (A:T, G:C), implied by the double-helix structure, requires the two strands of the double helix to be held together by hydrogen bonds. Thus, each strand of DNA is different: the nucleotides are lined up along the strand, being complementary to the nucleotides along the other strand and each strand having its own direction of propagation. The pairing of the bases is not necessarily perfect along the whole lenght of the double helix, and this leaves single-stranded regions which vary in length in the chain. The importance and the frequency of these single-stranded regions can be very different according to the origin (species and organ) of the DNA. DNA rich in G:C pairs (three hydrogen bonds) is more resistant to thermal denaturation than DNA rich in A:T pairs (two hydrogen bonds). The hydrogen bonds make the molecule both flexible and rigid.

All dividing cells contain DNA polymerase(s). This enzyme catalyzes the formation of DNA from deoxyribonucleoside-5′-triphosphates, using single-stranded DNA as template. It carries out the replication to produce complementary chains which are the exact replicas of each strand of the DNA. Thus, the separation of the two strands of the double chain is an essential operation. It proceeds as the DNA polymerase accomplishes the synthesis of the complementary chains. In the semi-conservative replication model [*Meselson and Stahl*, 1958], each strand of DNA serves as a template for the positioning of the deoxyribonucleotides. They line up sequentially and respect the complementarity of the bases (hydrogen bonds) on the original strand which, with the new strand, forms a double-helical chain of DNA. The nucleotides of the new chain are bound together by phosphodiester linkages [*Watson and Crick*, 1953]. This reaction requires the presence of Mg^{++} or Mn^{++}. Each of the four deoxyribonucleoside-5′-triphosphates is degraded into nucleoside monophosphate and pyrophosphate. Deoxyribonucleoside-5′-triphosphates produce the energy necessary for the initiation of the reaction, and each nucleotide carries the 3′-OH group which binds the next nucleotide. The deoxyribonucleotides polymerize sequentially at the replication fork (replication regions) [*Gefter*, 1974; *Dressler*, 1975; *Liu* et al., 1978]. Each strand of newly synthesized DNA is antiparallel and has one terminal 3′-OH group and one 5′-OH (or 5′-P or 5′-tri-P). The replication of DNA progresses along the chromosome in the direction of 5′ to 3′. The polymerization of the nucleotides may be continuous on strand 3′-OH (iso) following the progression of the replication forks, but on strand 5′- (anti) the replication seems to be formed from Okazaki fragments which are fragments of DNA synthesized discontinuously in the direction of 5′ to 3′ [*Okazaki* et al., 1968, 1969; *Lehman* et al., 1958; *Lark*, 1972]. The fragments of DNA (1,000–2,000 nucleotides) can be bound together in a long chain by polynucleotide ligase. This linking enzyme [*Lehman*, 1974] first activates the 5′-terminal group of the DNA template by fixing one AMP molecule, then provokes a nucleophilic attack of the adjacent 3′-OH which liberates the AMP.

Different DNA Polymerases

With some difficulty, DNA-dependent polymerases were isolated from bacteria [*Kornberg*, 1974a; *Gefter*, 1975], eukaryotic cells [*Bollum*, 1960; *Weissbach* et al., 1975; *Chang and Bollum*, 1971; *Chang*, 1973; Barry and Gorski, 1971] and mitochondria [*Kalf and Ch'ih*, 1968; *Meyer and Simpson*, 1970]. The three bacterial enzymes I, II, III and the three eukaryotic enzymes α [*Bollum*, 1960]; β [*Weissbach* et al., 1971; *Chang and Bollum*, 1972; *Chang*, 1975; *Barry and Gorski*, 1971] and γ [*Weissbach* et al., 1975; *Lewis* et al., 1974] present three essential differences: their molecular weights, their sensitivities to blocking agents of the sulfhydryl group, and their abilities to transcribe DNA or oligopolymers of AnT_{15} [*Bollum*, 1960; *Chang and Bollum*, 1971; *Chang*, 1973]. The α and γ polymerases are acid proteins and have molecular weights greater than 100,000. The β polymerases are alkaline proteins and have molecular weights less than 50,000. The purification of these proteins is difficult and so is the determination of each unit in the enzyme. One DNA polymerase with a molecular weight of 27,000 and without subunits has been isolated and purified from chicken embryos [*Stravianopoulos* et al., 1972].

DNA polymerase β, which replicates the double chain of 'activated DNA' [*Aposhian and Kornberg*, 1962] but which only weakly copies the oligopolymer AnT_{15} [*Bollum*, 1960; *Weissbach* et al., 1971], is generally found in cytoplasmic extracts but it may also be present in the nucleus. DNA polymerase β is very active during the S period of the cellular cycle, i.e., the period of active synthesis of DNA in vertebrates. This activity is negligible during the other periods of the cellular cycle [*Keir* et al., 1977]. DNA polymerase α, concentrated in the nucleus [*Weissbach* et al., 1971; *Chang and Bollum,* 1971; *Chang,* 1973], replicates activated DNA, whereas DNA-dependent DNA polymerase γ copies the oligo-homopolymers of AnT_{15} efficiently but DNA poorly [*Fridlender* et al., 1972; *Lewis* et al., 1974]. In spite of numerous studies, the specific function of each DNA polymerase is not yet clearly known.

Opening of the Double Helix and Replication of the DNA Strands

Many experiments have shown that as soon as the double helix starts to unwind, the separation of the two strands begins and DNA polymerase catalyzes DNA synthesis. Among numerous factors, certain proteins favor the separation of the strands as the enzyme progresses along its template. The bidirectional nature and opposed directions of replication prevent the rapid opening of the helical structure. The DNA double helix is not a static structure. The rupture of the hydrogen bonds between the two strands can be brought about by an increase in pH change, in temperature, by a reduction of ionic strength, etc. DNA polymerase itself may have a weak effect on strand separation, but this effect is not sufficient by itself. Hence, studies have been carried out to identify the proteins which, although they cannot polymerize deoxyribonucleotides, can open the DNA double helix (fig. 1). Proteins of this kind, known as unwinding enzymes [*Herrick and Albertz,* 1976; *Herrick* et al., 1976]. Untwisting enzymes [*Molineux* et al., 1974; *Yarrouton* et al., 1979; *McPherson* et al., 1979; *Curmingham* et al., 1979], gyrases [*Gellert* et al., 1979], are involved in the removal of supercoils in the DNA, supercoils which form in DNA ahead of the replication fork. ATP-dependent unwinding enzymes which provoke strand separation are also known [*Abdel-Monem and Hoffman-Berling,* 1980]. DNA unwinding enzymes have been isolated from eukaryotic organisms including mammals, amphibia and insects [*Champoux,* 1978]. It has been shown that nucleases, ATPases [*Sigal* et al., 1972; *Weiner* et al., 1975], ligases [*Bertazzoni* et al., 1971, 1976; *Lehman,* 1974], also known as nicking-closing enzymes, are involved in DNA replication. However, the activities of all these enzymes cannot yet be coordinated with those of the DNA polymerase. Their function may be not only to separate the two strands but to maintain them in the single-stranded state, thus favoring the progression of DNA polymerase.

Considering the problems of steric crowding posed by the action of this large number of molecules, it seems possible to ask whether some of these

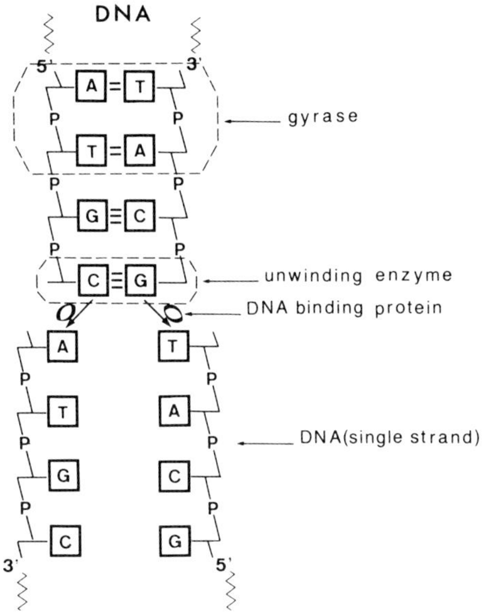

Fig. 1. Schematic representation of DNA strand separation.

activities may not be pursued by the same molecules. However, whatever model of DNA replication is considered, opening the double helix is a basic necessity [*Watson and Crick,* 1953], and it seems that a DNA polymerase recognizes only the sugar phosphate part of nucleotides and therefore requires a single strand as template and RNA primer for the synthesis of complementary sequences. The principle of semi-conservative replication of DNA has been solidly established [*Meselson and Stahl,* 1958].

Initiation of DNA Synthesis

Numerous experiments have shown that none of the known DNA polymerases can initiate the replication of a single intact DNA strand in vitro. In other words, DNA polymerase cannot create phosphodiester links between deoxyribonucleotides in the presence of a DNA template unless a free 3′-OH residue, which may be either ribo- or deoxyribonucleotide, is present to permit initiation of replication. Partial opening of the DNA double helix implies the existence of multiple initiation signals [*Haskell and Danern,* 1969; *Brewer,*

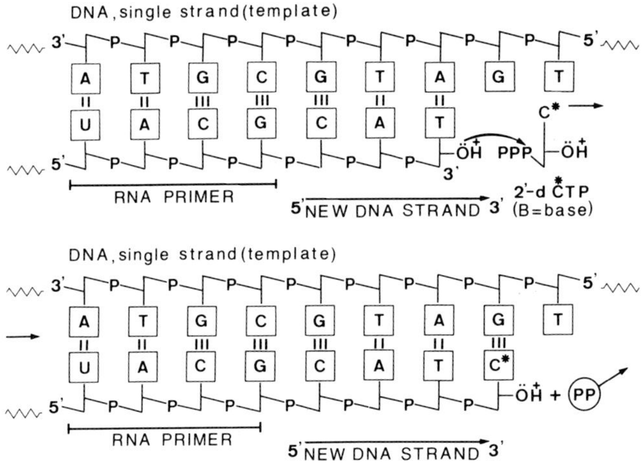

Fig. 2. Schematic representation of DNA replication by DNA-dependent DNA polymerase.

1972]. These have in fact been found in Chinese hamster cells [*Amaldi* et al., 1972; *Huberman and Riggs,* 1968] and in HeLa cells [*Cairns,* 1966]. Bubbles in the DNA, corresponding to replication zones, have been observed by electron microscopy [*Delius* et al., 1971; *Bourgaux and Bourgaux-Ramoisy,* 1972]. Nascent DNA strands in these strands may correspond to *Okazaki's* DNA fragments (8–10S). This problem is extensively discussed in a general model of DNA replication [*Ioannou,* 1973], in which bidirectional replication in the bubbles is also described.

The nucleotides carrying a free 3′-OH group can be provided by single-strand cleavages in a DNA molecule or by the 3′-OH terminus of an oligonucleotide. In addition, DNA-dependent RNA polymerase can synthesize a short RNA chain which can function as an initiator of DNA replication [*Stravianopoulos* et al., 1971]. Starting from the free 3′-OH, DNA polymerase can attach the first deoxyribonucleotide to the chain which is bound to the template by hydrogen bonds (fig. 2). In the present state of knowledge, it is not possible to exclude any of the possible mechanisms of production of the free 3′-OH groups necessary for the initiation of the replication. Fragmentation of DNA by a nuclease is probably not specific and might be expected to lead to random or incomplete replication. This type of activation of DNA would resemble that produced by low doses of X rays.

The initiator RNAs, which contain between 15 and 50 ribonucleotides, form a family of RNAs each of which might initiate the replication of a specific unit of DNA (operon, unitron or several neighboring operons). The existence of a covalent bond between the 3′-OH group of the RNA initiator and the first deoxyribonucleotide of the nascent DNA chain has been demonstrated [*Pigiet* et al., 1974; *Okazaki* et al., 1975; *Blinkerd and Toliver,* 1974; *Blair* et al., 1972; *Williams* et al., 1973; *Wagar and Huberman,* 1975].

RNA Primers

RNA polymerase has two essential properties which allow one to visualize its role in the synthesis of RNA primers: when bound to double-stranded DNA, this enzyme provokes the separation of DNA strands over short distances. RNA polymerase catalyzes the polymerization of all four ribonucleotides (from ribonucleoside-5′-triphosphates) by positioning them on only one single strand of the DNA. Experiments performed in vitro using either relatively purified enzymes or cellular extracts have shown that the synthesis of new DNA chains cannot be achieved either in the absence of RNA polymerase or in the absence of the four ribonucleoside-5′-triphosphates [*Chargaff,* 1977; *Blair* et al., 1972].

The first demonstration that an RNA primer acts directly as a primer in the initiation of DNA replication [*Stravianopoulos* et al., 1971] was made when using either a purified DNA polymerase from chicken embryos (the molecular weight of this enzyme is 27,000 and contains no subunits) or a crude cellular extract. Results obtained with these two systems are practically identical. DNA polymerase behaves differently depending on the template DNA, intact or denatured (pretreatment with 0.1 *N* KOH for 20 min at 60 °C), or 'activated' by a small amount of DNase which causes nicks [*Goulian,* 1969]. The velocity and quantity of synthesized DNA is thus dependent on the integrity of DNA: intact or denatured DNA (the latter has completely separated strands) is not very active as template. On the contrary, 'activated' DNA is an excellent template. Single-stranded DNA (denatured DNA), however, may become an excellent template, provided the reaction medium contains the RNA polymerase and all four ribonucleoside-5′-triphosphates which are the conditions required for the synthesis of RNA primers [*Beljanski* et al., 1975; *Chargaff,* 1977]. In this latter research work, it has been shown that synthetic polymers, poly-r(A), poly-d(T), are used as primers for DNA synthesis. In this reaction, the concentration of Mn^{2+} ions determines not only the amount of synthesized RNA but also the number of initiation points [*Stravianopoulos* et al., 1972; *Chargaff,* 1977]. Thus, three points of initiation have been found for AMP and five for GMP per DNA template molecule. When RNA polymerase activity was inhibited by rifampicin or by chloramphenicol, an RNA primer could not be synthesized by RNA polymerase, thus making it impossible for the DNA-dependent DNA polymerase to synthesize DNA. Recently, it was reported that mixed primers (ribo- or deoxyribonucleotide oligomers) [*Roven and Kornberg,* 1978; *Benz* et al., 1980] might serve as starting points for DNA synthesis. Such results have been obtained by different authors, using either bacterial lysates [*Schekman* et al., 1972; *Okazaki* et al., 1975; *Plawecki and Beljanski,* 1974; *Beljanski* et al., 1975; *Hirose* et al., 1973] or lysates of animal cells [*Pigiet* et al., 1974; *Brun and Weissbach,* 1978]. The possibility that DNA replication might be initiated by RNA primers is no more doubtful. The RNA primer has to be localized with precision on DNA in a site where complementarity of bases between DNA and RNA primers might take place. A normal cell must necessarily avoid incorrect initiation of replication which might be a source of errors in transcription. Thus, in vitro DNA polymerase accepts different oligonucleotides (mostly oligoribonucleotides) as primers [*Wells* et al., 1972; *Chang and Bollum,* 1972; *Keller,* 1972; *Beljanski and Aaron-da Cunha,* 1976; *Dressler,* 1975], even DNA primers [*Gefter,* 1974]. However, one should make sure that each of these primers might be utilized in vivo for the initiation of DNA synthesis occurring in an extremely complex environment. Therefore, in case of error, RNA primer should be more easily eliminated than DNA primer [*Stravianopoulos* et al., 1972]. In fact, observations reported elsewhere show that priming by RNA is a current and naturally occurring process, both in vitro and in vivo. DNA priming by a DNA primer remains a rather exceptional event.

The role of the RNA primer is solely to initiate a given site of replication on DNA. It is neither transcribed nor incorporated into DNA. Once the initiation process has been accomplished, the RNA primer is eliminated by neighboring nucleases. Resulting nucleotides return into a pool of biochemical pathways involving synthesis of polyribonucleotides. RNA primers do not represent any danger as far as the integrity of the genome is concerned. We shall illustrate this point in chapters to come.

The availability and nature of primers present in cells should favor the replication of different segments of DNA and interfere in the alternation between replication and transcription. Depending on their nature (amount of purine and pyrimidine bases) and affinity between RNA primer and a segment of DNA, the replication may precede transcription or vice versa. Thus in fact, during the differentiation of cells, there must be priority, even if only temporarily, of transcription over replication.

RNA Primers and Replication of Phage Circular DNA

Linear DNAs, formed from two complementary strands, are not the only DNAs that require the presence of RNA primers for replication. In fact, circular single-stranded DNA also requires 'the primers'. Using a single circular strand of bacteriophage F_1DNA, RNA polymerase synthesizes an RNA in vitro, permitting the DNA polymerase to start the replication of DNA [*Stravianopoulos* et al., 1971]. Such synthesized RNA contains about 30 ribonucleotides whose sequences were determined beforehand [*Geider* et al., 1978]: the quantity of purine bases (A+G) is twice that of pyrimidine (C+U). The synthesis of phage DNA occurs neither in the absence of RNA polymerase nor in the absence of nucleotide precursors for RNA. When using rifampicin, which inhibits the activity of RNA polymerase, synthesis of RNA is prevented and, consequently, synthesis of DNA is considerably slowed down [*Schekman* et al., 1972; *Chargaff*, 1977]. If rifampicin is added to the incubation mixture after the reaction for RNA synthesis has started, a normal quantity of DNA is synthesized. Partly purified from *E. coli* extracts, DNA-dependent DNA polymerase cannot synthesize in vitro the DNA of phages ΦX174 or lambda [*Beljanski*, 1975] in the absence of RNA polymerase or specific RNA primers. The initiation site of ΦX174 phage DNA contains, almost exclusively, pyrimidine bases [*Ling*, 1972], and thus primers with low G and A contents are poorly active in this system (no complementarity of bases). On the other hand, in the presence of exogenously prepared RNA primers rich in G and A nucleotides, the synthesis of both ΦX174 and λ DNA phages is highly increased, resulting in the formation of double-stranded DNA [*Beljanski*, 1975]. We have also shown that, in this reaction, hydrogen bonds are formed between phage DNA and the A-G-rich RNA primer. Deoxyribonucleotides incorporated into the newly synthesized circular strand show the same ratio as those found in the template DNA: the complementary strand has been well synthesized under RNA primers by DNA polymerase; even in the presence of rifampicin, exogenously prepared purine-rich RNA primers allowed the initiation of single-stranded circular DNA, replicated by DNA-dependent DNA polymerase I from *E. coli*. It was shown that in the T_4 bacteriophage, two proteins catalyze the synthesis of a group of pentaribonucleotides acting as primers for de novo DNA chain initiation on the lagging strand in the T_4 DNA replication [*Liu and Alberts,* 1981].

These data suggested that, in vivo, the exogenous RNA primers might stimulate the replication of DNA from rapidly dividing cells, such as hematopoietic cells, implying that synthesis of RNA primers precedes that of DNA. The discovery of oligoribonucleotides (10–100 nucleotides) in DNAs isolated and purified from bacteria [*Okazaki* et al., 1968], mammalian cells [*Pigiet* et al., 1974; *Brun and Weissbach,* 1978; *Fox* et al., 1973; *Tseng and Goulian,* 1977], from viruses [*Pigiet* et al., 1974; *Beljanski,* 1975], from plasmids [*Helinski,* 1976; *Blair* et al., 1972; *Williams* et al., 1973] would appear to indicate a general phenomenon in DNA replication. RNA primers have also been found at the origin of the mitochondrial DNA replication site (1–10 ribonucleotides dispersed on initiation site; *Gillum and Clayton,* 1979). Certain segments of DNA isolated from intracellularly developed phage contain D loops [*Kasamatsu and Vinograd,* 1974; *Brown* et al., 1978]. Once one of the two strands has been separated, it gives rise to loops (about 300 bases) where RNA was found to be present [*Chattoraj and Stahl,* 1980]. The possible role of these loops would be to participate in the replication and transcription of DNA [*Maitra and Hurwitz,* 1965]. However, this hypothesis will have to be confirmed by more direct experimental evidence.

RNA Primers and Discontinuous DNA Synthesis in vivo

The first experimental evidence of discontinuous in vivo synthesis of DNA requiring RNA primers has been reported by *Okazaki* et al. [1968]. They have shown that labelled thymidine is first incorporated into DNA fragments containing 1,000–2,000 nucleotides. Later on, these fragments have been found incorporated into a long DNA chain. In vivo, such DNA fragments might contain 80–120 ribonucleotides. The size of these RNA chains appears too high to play a role as RNA primers. Usually, RNA primers contain 10–50 nucleotides.

During polyoma DNA synthesis, 'Okazaki DNA fragments' in isolated nuclei were found to contain RNA primers (10 ribonucleotides) associated with the 5′-terminus of 4S DNA pieces [*Pigiet* et al., 1974; *Brun and Weissbach,* 1978]. Also, in a system of HeLa cells, discontinuous DNA synthesis depends essentially on the presence of all four ribonucleoside-5′-triphosphates which are the nucleotides required for RNA synthesis. Omission of one of the four ribonucleotides leads to a very low level synthesis of 4S DNA fragments. Various data suggest that once the RNA primer action has been accomplished, it is removed from the DNA chain which is then completed by DNA polymerase and finally ligated by polynucleotide ligase [*Lehman*, 1974] in order to form a continuous complete chain. Ribonuclease H [*Keller and Crouch*, 1972] or other types of nucleases present in DNA polymerase preparations might have the task of eliminating the RNA primer once its function has been terminated.

The fact that the activity of RNA polymerases II and III is inhibited by α-amanitin, while the synthesis of 4S DNA pieces in the presence of four ribonucleoside-5′-triphosphates is completely insensitive to this inhibitor, suggests that RNA polymerase I is implicated in the synthesis of RNA primers [*Stravianopoulos* et al., 1971; *Brun and Weissbach*, 1978]. However, one should keep in mind that the control of DNA synthesis is very complex, especially in vivo, and that other undefined molecules or factors are involved. We do not know how the coordination of the activities of different enzymes, factors and histones can be regulated on the biochemical level [*Bryant*, 1980], although DNA replication is regulated biochemically, physiologically and morphologically.

Exogenously Prepared RNA Primers and DNA Replication

In eukaryotic cells, RNA primers may originate, for example, during transcription of DNA into RNA, during degradation of ribosomal RNAs or from precursors of mRNA or during degradation of intervening sequences in the mRNA by means of specific nucleases. We have made great efforts to obtain large amounts of RNA primers which differ according to size and base composition.

Mild in vitro degradation of purified ribosomal RNAs originating from bacteria or mammalian cells using different ribonucleases (pancreatic RNase, RNase T_1 and U_2 etc.) brings about the appearance of families of RNA fragments containing 20–65 ribonucleotides [*Beljanski* et al., 1975]. The presence of nucleotide-3′-phosphatase activity in all RNases used confers on most RNA fragments a free 3′-OH group required for their priming activity. Long-chain RNA degradation allowed us to get large amounts of a great variety of RNA fragments which were selected according to size. They have different ratios of purines and pyrimidines and contain a different 3′-OH terminal nucleotide [*Beljanski* et al., 1975]. In this way, it appeared possible to perform several types of experiments in order to determine the biochemical and biological in vitro and in vivo capacities of these RNA fragments.

Before examining different examples in which exogenously prepared RNA fragments have been used, it is essential to consider the fact that RNA fragments obtained by using RNase A or T_1 are not transcribed in vitro into DNA by a reverse transcriptase-like enzyme [*Beljanski and Aaron-da Cunha,* 1976; *Beljanski* et al., 1975; *Le Goff and Beljanski,* 1979]. This is an extremely important observation since, if used in vivo, RNA fragments must be harmless for the integrity of the genome [*Beljanski* et al., 1978c, 1981a]. RNA primers have one essential property, i.e., to act as an initiator for DNA replication.

Specificity of RNA Primers; Multiple Initiation Sites

To demonstrate the effect of RNA primers, single- or double-stranded DNA samples have to be carefully purified in order to contain only a very limited amount of contaminants (proteins 0.5%, RNA 5%). Complete elimination of RNA as a contaminant in DNA can be achieved by incubating DNA samples with alkali (0.3 *M* KOH, 24 h at 37 °C), followed by an extensive dialysis. Such DNA, free from RNA contaminant, will readily react with adequate primers. DNA-dependent DNA polymerase I has also to be purified in such a way as to exclude RNA primer types. If the DNA used contains RNA, addition of RNA primers to the incubation mixture results in a 7- to 10-fold increase in DNA synthesis. RNA contaminant might have no biological activity and may prevent some segments of DNA from being replicated. When both DNA and enzyme are practically free of RNA, the stimulating effect of exogenous RNA primers for DNA synthesis in vitro may be of the magnitude of 50-fold. Neither intact ribosomal RNAs nor transfer RNAs can replace active RNA primers containing 20–50 ribonucleotides [*Beljanski,* 1975].

A family of RNA fragments differing in size and base composition has been obtained by the degradation of purified ribosomal RNAs of *E. coli* M 500 Sho-R strain with pancreatic RNase or other RNases under well-defined conditions [*Beljanski,* 1975]. The advantage of using rRNA from this strain is that both ribosomal RNAs (23S and 16S) contain twice as much purine than pyrimidine nucleotides [*Beljanski and Beljanski,* 1968; *Beljanski* et al., 1972b]. The resulting RNA fragments, fractionated on Sephadex G-25 column, show a rather broad specificity when tested for the in vitro synthesis of various DNAs [*Beljanski,* 1975]. In some cases, they are quite highly

Table I. Replication of DNA in vitro from various tissues in the absence or presence of RNA fragments [a]

Incubation mixture	^{3}H-TTP [b] incorporated into DNA, cpm								
	10 min			20 min			30 min		
	DNA 1	DNA 2	DNA 3	DNA 1	DNA 2	DNA 3	DNA 1	DNA 2	DNA 3
Complete	550	451	466	662	520	507	701	580	660
+RNA fragment (4 μg)	1,980	1,276	506	2,164	1,476	617	2,256	1,634	756
+RNA fragment (4 μg) +DNase (2 μg)	112	104	110	123	126	98	130	–	–
+RNA fragment (4 μg) – DNA	102	107	98	133	112	108	140	–	132

[a] The incubation medium contains per 0.15 ml: Tris-HCl buffer, pH 7.65: 25 μmol; $MgCl_2$: 2 μmol; 4 *d*-XTP (deoxyribonucleoside-5′-triphosphates): each 5 nmol (+ ^{3}H-TTP: 50,000 cpm); DNA: 0.2 μg; RNA fragments: 4 μg; DNA-dependent DNA polymerase: 80 μg. Incubation was for 10, 20, and 30 min at 36 °C. Trichloroacetic acid-precipitable material was filtered on GF/C glass filter, washed, dried, and the radioactivity was measured in a Packard liquid spectrometer. DNA 1 = Bone marrow; DNA 2 = spleen; DNA 3 = brain.

[b] TTP = Thymidine-5′-triphosphate.

specific. The specificity exhibited by RNA primers can be observed with DNA from chosen species categories (viruses, phages, bacteria, plants, mammals) or with DNA originating from different organ tissues of a given animal. Thus double- and single-stranded DNA, originating from various organs of an animal inside a given taxonomical species, respond differently to RNA primer during synthesis. Sometimes a very high specificity may exist between the RNA primer and DNA (table I). It emerged therefore that the nucleotide composition of an RNA primer is more important than its size regarding the capacity of the primer to bind to a given DNA. This is necessarily due to the need for complementarity between the RNA primers and the segments of DNA on which the initiation sites required for replication are located. The RNA primer active in the replication of single-stranded circular DNA of phage ΦX174 contains a large excess of G and A nucleotides [*Beljanski,* 1975]

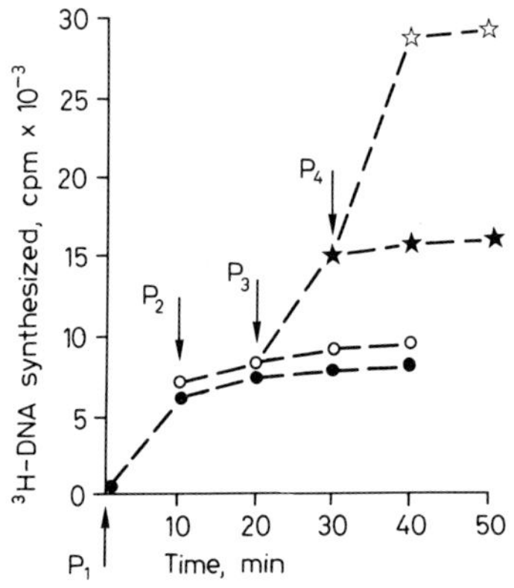

Fig. 3. Effect of various exogenous RNA fragments (primers) on quail liver DNA synthesis in vitro [from *Beljanski* et al., 1975].

which suggested that the initiation site for DNA replication phould contain pyrimidine bases. In fact, sequence analysis of the initiation site on DNA ΦX174 phage had shown the almost exclusive presence of C and T nucleotides [*Ling*, 1972].

For a given amount of DNA, the absence of competition between some RNA primers, each utilized at saturating concentrations, suggests several initiation sites for DNA replication. Thus, DNA from quail liver responds in vitro differently to various RNA fragments (RNA primers) obtained in the presence of pancreatic RNase. At saturating concentrations, RNA primers P_1 and P_2, both rich in G and A nucleotides, do not compete for a given initiation site or sites on DNA. After the synthesis of this DNA has been maximally stimulated by $P_1 + P_2$ RNA primers, the addition of another RNA primer P_3 at saturation level raised the synthesis of DNA. This may be further increased by a P_4 RNA primer (fig. 3). Thus the same DNA provides different sites for different RNA primers responsible for different levels of synthesis on the same DNA template. These different levels would correspond to different segments in the DNA to be replicated. It will be of interest to determine if a given RNA primer initiates the replication of a given genetic unit [gene, operon(s)]. This appears probable, especially if DNA undergoes arrhythmic replication, which seems to be the case [*Nagl and Rücker*, 1976]. Does the RNA primer give the signal only at the initiation site on DNA, punctuating the secondary structure on the spot where discontinuous synthesis of

DNA takes place? If, as seems likely from the above results, different RNA primers initiate the replication of different portions of DNA, they should contribute to the flexibility of DNA during the differentiation of cells. The diversity of RNA primers may also contribute in protecting the replication sites on DNA from incorrect and unforeseeable RNA primers. It is important to determine whether DNA itself selects primers or whether DNA-dependent DNA polymerase participates in this selection. From our own results, both seem to be required and this might have as a consequence an increase in the rate of DNA strand separation. The portion of DNA to be chosen would depend on both RNA primer and DNA polymerase, which process does not exclude the participation of other types of molecules.

Pretreatment in vitro of RNA primers by both RNase P and T_1 leads to a considerable decrease in their priming effect in DNA synthesis, as was shown when single-stranded DNA from ΦX174 phage was used as template [*Beljanski*, 1975]. In vivo, RNase H associated with DNA polymerase may contribute in eliminating the RNA primers [*Keller*, 1972] or adjusting the size of a primer according to the need of a cell. This enzyme may also participate in removing undesirable RNA primers, thus strengthening the security of the whole process of replication.

The theoretical model [*Ioannou*, 1973] implies the formation of a covalent linkage between RNA primer and the very first deoxyribonucleotide to which the next nucleotide will attach. Formation of hydrogen bonding between RNA primer and a segment of DNA has been confirmed by our experiments in which exogenous RNA primer was used for DNA synthesis [*Beljanski*, 1975]. In fact, inactivation of the 3′-OH group of the RNA primer with periodate and the formation of covalent linkage between RNA primer and the newly synthesized DNA chain (only if active DNA synthesis is promoted in the system) are in complete agreement with the proposed model.

RNA Primers and the Genesis of Leukocytes and Platelets in situ

Results obtained with exogenously prepared RNA primers suggested that some of them could initiate DNA replication in vivo, thus permitting cells to divide [*Beljanski* et al., 1978c, 1981a]. In this research we have chosen to follow the genesis of hematopoietic cells which rapidly divide in animals and humans. The choice for RNA fragments was based on three main observations:

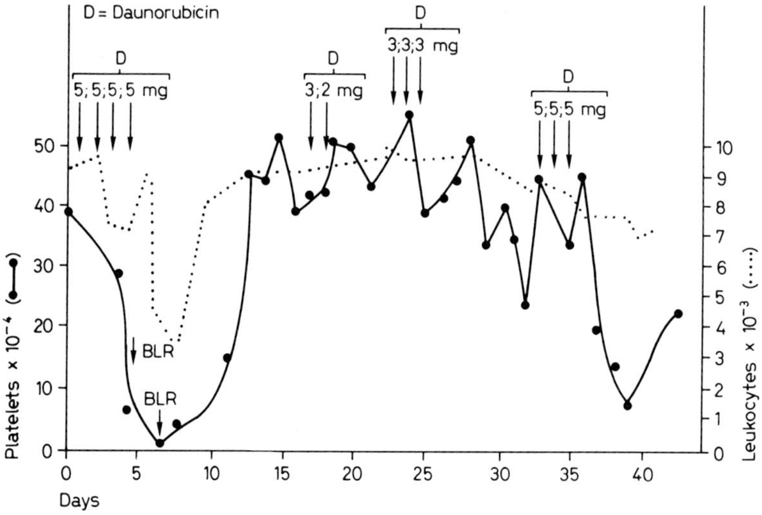

Fig. 4. Effect of BLR (Beljanski Leukocyte Restorer) on leukocyte and platelet counts in the rabbit pretreated with high dosages of daunorubicin. A 4-kg rabbit received intravenously 5 mg daunorubicin daily for 4 consecutive days. At the time shown by arrows the rabbit received two BLR dosages (5 mg i.v. and 20 mg per os). The leukocyte count, which was low in the drug-treated animal, was brought back to 10,000 within 48 h. After treating with BLR, several doses of daunorubicin were injected intravenously (arrows). The same experiment was repeated on another rabbit. Platelet counts were performed for both animals. 3 control rabbits treated only with daunorubicin died at the 6th to 9th day following the first daunorubicin injection [from *Beljanski* et al., 1981a].

(a) a particular family of RNA primers strongly stimulates the in vitro synthesis of DNA purified from rabbit bone marrow and spleen while they have practically no effect on the synthesis of DNA from other tissues;

(b) these RNA primers are not transcribed into DNA, thus eliminating the risk that the genes might be altered;

(c) they localize in hematopoietic tissues where, considering their excellent priming activity on DNA, they were expected to induce leukopoiesis in animals having a medullary aplasia. They have no toxic effect on the organism to which they are administered.

When administered intravenously, intramuscularly, intraperitoneally or even orally to rabbits, RNA primers localize in the bone marrow and spleen. They rapidly restore a normal level of the leukocyte and platelet count whose levels had been drastically decreased either by chemotherapy or radiation.

During this process, one may observe the reestablishment of the normal balance between granulocytes and lymphocytes which has been upset by various chemical drugs or which may have appeared for unknown reasons. Frequent intravenous injections of RNA primers do not lead to a tolerance phenomenon nor to any detectable toxicity [*Plawecki and Beljanski,* 1981].

Experimental evidence shows that, in animals, RNA primers induce differentiation of stem cells into granulocytes, lymphocytes and platelets within a rather short period of time. It should be emphasized that animals to which an excess of RNA primers active in leukopoiesis was injected show no particular response [*Beljanski* et al., 1981a]. There is no effect of RNA primers on cancerous cells or pathological leukocytes in either direction. It was remarkable to observe that the very powerful stimulating effect on DNA from hematopoietic cells occurs only with DNA from healthy cells. It appears possible and even probable that genesis of pathological cells needs other types of RNA primers, differing in base composition and size from those used by healthy cells. On this basis, one would expect that in cancerous cells RNases should exhibit quantitatively modified properties, as has already been shown for RNase present in plant tumor cells [*Reddi,* 1966]. A second conclusion from these studies suggests a close specific relationship between the activities of RNA primer and RNase(s) in a given tissue, particularly in those which undergo cell division. RNA primers involved in leukopoiesis and platelet formation (fig. 4) [*Beljanski* et al., 1981a] have been successfully used in treating human medullary aplasia which appears during and after chemotherapy of cancers.

Ribosomal RNAs as Source of RNA Primers

Our in vitro and in vivo results have shown that ribosomal RNAs are a potential source for the generation of active RNA primers if they undergo degradation in the presence of a given RNase. Taking into account the data described by various authors and those described above, it may be thought that such events must operate in vivo. That ribosomal RNAs (or precursors of these RNAs) might be involved in generating RNA primers is supported by the following observations:

(a) The amount of RNA synthesized in the nucleolus is 10 times higher than that observed in the nucleus [*Yu,* 1980]. An RNase present in the nucleolus participates in the processing of functional ribosomal RNAs from their giant precursors [*Grummt* et al., 1979].

(b) Hormones strongly stimulate the synthesis of ribosomal RNAs in the nucleolus and that of other RNAs in the nucleus [*Luck and Hamilton,* 1972; *Hiremath and Wang,* 1979] of eukaryotic and plant cells.

(c) In eukaryotic cells a large number of ribosomal RNA copies are produced during the first stage of development (before cell multiplication), and they disappear thereafter [*Brown and Tocchini-Valentini,* 1972; *Cullis and Davies,* 1975]. Conidia cells of *Neurospora crassa* in germination contain 250–300 more copies of ribosomal RNAs than mycelia cells. The extra copies of ribosomal RNAs disappear as soon as cell multiplication starts [*Mukhopadhyay and Dutta,* 1979]. It has been suggested that gene amplification of ribosomal RNA could proceed through a 'chromosome mechanism' [*Brown and Blackler,* 1972] in which RNA-dependent DNA polymerase (reverse transcriptase) might play a crucial role. This enzyme has also been found in *Neurospora crassa [Dutta* et al., 1977].

Environmental conditions can induce the increase of ribosomal RNA copies [*Durrant,* 1962]. Thus, the taller forms of the flax plant grown in the presence of certain fertilizers contain nuclei in which the amount of DNA increases as well as the size of the plant nucleoli. This phenomenon might reflect the activity of redundant ribosomal genes, as was shown for the axolotl [*Miller and Brown,* 1969]. In fact, hybridization techniques have shown that, after fertilization, the high forms of the flax would contain 47–65% more copies of ribosomal RNA and their size increased four times when compared with initial plant size [*Timmis and Ingle,* 1973]. One may conclude that when exogenous chemical substances are implicated, the DNA segment corresponding to genes for ribosomal RNA undergoes strand separation, thus allowing many copies of ribosomal RNA to be made. Consequently, excess ribosomal RNAs may be degraded by RNases. This provides RNA primers which induce the replication of DNA, followed by cell division and development. These data suggest the existence of a relationship between the accumulation of ribosomal RNA copies made in excess at the beginning of the induction of cell multiplication and plant development. In addition, some such RNA fragments migrate into the cytoplasm, allowing the initiation or inhibition of the translation process.

Inhibitors of DNA Synthesis

Numerous agents which occur in nature or which may be chemically synthesized inhibit in vitro and in vivo DNA synthesis. Some induce errors in the polymerization of deoxyribonucleotides during the synthesis of DNA chains, while others bind with DNA, making a complex which prevents the DNA-dependent DNA polymerase from carrying out its function. Substances also exist which directly interact with enzyme molecules and thus strongly reduce DNA synthesis. This problem has been extensively discussed [*Loeb* et al., 1980], and we shall quote only a few examples concerning in vitro and in vivo synthesis of DNA in the presence of particular polyribonucleotides or antimitotic compounds.

Effect of Polyribonucleotides

The activity of DNA polymerase of leukemic viruses is diminished in the presence of single-stranded homopolymers [*Srivastava*, 1973; *Schrecker* et al., 1974]. Low concentrations of poly (U) inhibit viral DNA polymerase activity more efficiently than the activity of the same enzyme in eukaryotic cells. But α-polymerase appears to be an exception. In contrast, high concentrations of poly (U) inhibit in vitro the activity of both α- and β-polymerases [*Abrell* et al., 1972; *Tuonunen and Kenney*, 1971]. These results should have drawn more attention to the importance of small-size specific RNAs which appear during DNA transcription or RNA degradation in eukaryotic and plant cells.

Recently, we have shown that RNA fragments containing about 65 ribonucleotides, obtained by the degradation of *E. coli* ribosomal RNA with RNase U_2, interact in vitro and in vivo with cancer DNA originating from plant tumor cells and from animal cancer tissues. In vitro, these U_2-RNA fragments strongly inhibit the synthesis of DNA purified from plant crown-gall tumors and DNA isolated from various mammalian cancerous tissues [*Le Goff and Beljanski*, 1979]. They are without effect on the in vitro synthesis of DNA of healthy plant and animal tissues. These results demonstrate that U_2-RNA fragments do not interfere with enzyme activity but with DNA as template. In vivo, these RNA fragments provoke necrosis of crown-gall cells, which correlates with their inhibitory in vitro effect on cancer DNA synthesis (fig. 5).

U_2-RNA fragments have no toxic effect on plant normal cells (pea seedling in germination) [*Beljanski* et al., 1978a, b]. Intact ribosomal RNA cannot replace U_2-RNA fragments. In this connection, it should be recalled that dialysis of microsomal preparations originating from tumorous or embryonic cells releases small-size RNAs which act specifically against tumorous cells

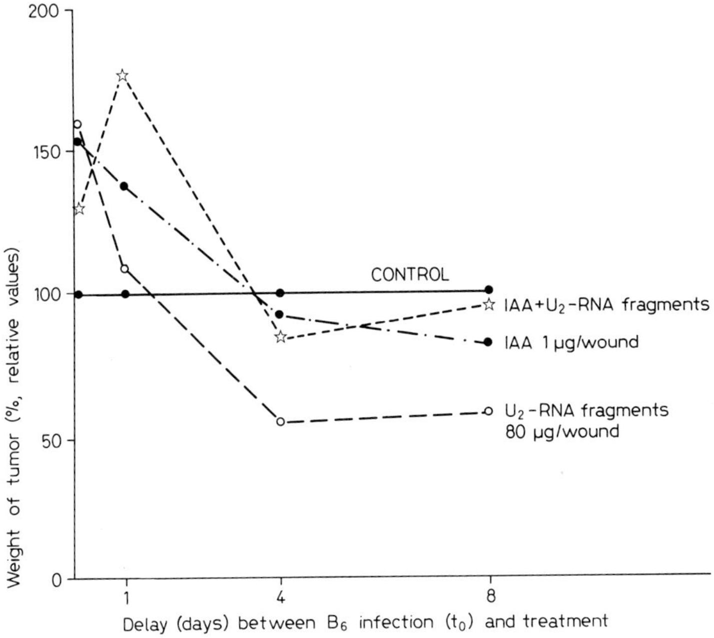

Fig. 5. Effect of U_2-RNA fragments on crown-gall tumor development. IAA = Indolylacetic acid [from *Le Goff and Beljanski,* 1979].

in vivo [*Pottathil and Meier,* 1977]. Similar results were obtained using the extracts containing the 'tumor-immune RNA' [*Schlager* et al., 1975]. For this reason they might be of importance during transcription and replication of DNA and possibly in cell differentiation. While U_2-RNA fragments act directly on DNA used as template, preventing in this way the activity of DNA polymerase, other molecules, such as specific antisera, act directly against the enzyme, and their inhibitory effect can be abolished by adding an excess of enzyme to the incubation mixture [*Aaronson* et al., 1971; *Loeb* et al., 1980].

Effect of Antimitotic Compounds on DNA Replication

The anthracyclins (daunorubicin, doxorubicin, rubidazone) intercalate between DNA strands and prevent the access of DNA-dependent DNA polymerase from fulfilling its function [*Dimarco* et al., 1971; *Gabbay* et al., 1976]. All three anthracyclins inhibit the activity of both α and β DNA polymerases with a preference for the α enzyme (enzymes were partially purified). DNA synthesis in isolated nuclei is also inhibited by these agents and thus corroborates results obtained with free enzymes [*Sartiano* et al., 1979]. We should recall that polymerase α is present in high amounts in dividing cells. It has been shown that anthracyclins increase the temperature of DNA denaturation, which concurs with the capacity of these agents to intercalate between DNA strands. However, it should be emphasized that in these experiments 'activated' DNA was used as template for DNA synthesis in the presence of high concentrations of anthracyclins, while for fusion experiments highly polymerized DNA was used. For this reason it is difficult to compare these two types of results.

Daunorubicin used in doses 100 times smaller than those mentioned above induces DNA strand separation, particularly in the case of cancer DNA. This results in a substantial increase in the DNA synthesis in vitro, while high concentrations are inhibitory [*Beljanski* et al., 1981a]. Other authors [*Tanaka and Yoshida,* 1980] have recently reported the inhibition of DNA polymerases α and β by daunorubicin and adriamycin which compete with RNA primers for DNA synthesis. Poly-d(T) and oligo-d(A) nucleotides act in such a way that the activity of DNA polymerase becomes much more sensitive to anthracyclins, as is the enzyme alone with the same concentrations of anthracyclins. The authors interpret these results by saying that drugs inhibit the activity of DNA polymerases. However, concentrations used are so high that it is difficult to interpret these observations correctly, although the preincubation of the drug with an enzyme (or enzymes) leads to a markedly decreased enzyme activity. Another compound, methotrexate, inhibits the synthesis of thymidylate [*Fridland and Brent,* 1975], and this results in an 80% decrease of DNA synthesis in human lymphoblasts cultured in vitro. Methotrexate inhibits the initiation of DNA chains, which results in a decreased amount of DNA synthesized independently of the growth of these cells [*Borsa and Whitmore,* 1969]. This drug also induces misincorporation of uracil into DNA [*Goulian* et al., 1980].

We have shown that actinomycin D very strongly inhibits the in vitro synthesis of DNA catalyzed by DNA-dependent DNA polymerase from *E. coli.* It acts less efficiently on the in vitro synthesis of RNA-dependent DNA polymerase [*Beljanski and Beljanski,* 1974]. This antibiotic binds to DNA by non-covalent linkage and does not allow access of DNA or RNA polymerase to DNA template. It is worthy of note that the activities of DNA polymerases α and β were decreased by 62 and 28%, respectively, 2 days after dexamethasone treatment of chick embryo, while DNA polymerase γ remained constant. The recovery was total at birth [*David* et al., 1980]. The disappearance of the 8.2S DNA ligase and the appearance of the 6.2S DNA ligase were observed earlier under the effect of corticosteroids in the thymus of the chick embryo. These results suggest that dexamethasone has allowed the expression of new genes connected with a decrease of the activity of some DNA polymerases.

Summary

We have assembled results showing the complex nature of various steps involved in the regulation of DNA replication. Single-stranded segments of DNA are required for DNA-dependent DNA polymerase(s) in order to link deoxyribonucleotides into a new complementary DNA chain. Untwisting and unwinding proteins (enzymes) induce DNA strand separation, provided there is a region of unpaired DNA. The synthesis of the new DNA chains requires enzymes and factors. However, we do not have knowledge how biochemical synthesis of DNA is regulated in a coordinative manner. DNA polymerases are not capable of initiating DNA replication without RNA primers although there is no certainty that both strands require RNA primers. Short-chain RNA primers bind to initiation sites on DNA, thus providing the 3′-OH group necessary for initiation. There exist numerous initiation sites on DNA in eukaryotic chromosomes. Replication is discontinuous, leading to the formation of Okazaki fragments which are thereafter joined by ligases. Each replication site may well be specific for a given RNA primer. Exogenously prepared RNA primers (they are not transcribed into DNA) induce in animals and humans the multiplication and differentiation of normal cells such as leukocytes and platelets, through initiation of DNA replication. RNA primers do not alter genes and the cell structure. Some observations indicate that RNA primers required by normal cells differ from those used by cancerous cells, since the former do not initiate the replication of DNA from cancerous cells. The appearance of RNA primers for DNA replication may be regulated by specific nucleases acting on various non-used RNAs or multiple redundant copies of ribosomal RNAs which disappear in a variety of cells as soon as cell growth starts. RNA fragments, endogenously formed or exogenously prepared, have the capacity to initiate or stop DNA replication and, consequently, cell division.

2. RNA Polymerases and Release of Information from DNA

Introduction

DNA carrying genes for structural and functional proteins of cells has to be transcribed into numerous RNA species in order to permit protein biosynthesis. RNA polymerases are enzymes which accomplish the transcription of DNA into several ribosomal and many messenger RNAs which are translated in the cytoplasm.[1] The role of these enzymes appears to be essential, and one cannot speak of the problems of the activity and differentiation of cells without keeping in mind how RNA polymerases bring about the release of specific information from DNA.

Here we shall outline some complex aspects concerning RNA polymerases and stress those aspects which are needed for a critical understanding of a scheme for gene activation which we intend to put forward in chapter 6.

Bases of deoxyribonucleotides contained in sequences of double-stranded helical DNA molecules contain information which determines the primary structure of proteins. Genes located inside DNA and responsible for constant and specific characters must be correctly transcribed into messenger RNA (mRNA) with respect to complementarity of bases. This process is called transcription. Later the mRNAs are translated into proteins.

Transcription of DNA into mRNA is an essential step in the functioning of cells, both dividing and 'resting', and also of cells which are in the process of differentiation. Although simple at first sight, this process is in fact complicated and difficult to grasp in all its details, especially during differentiation. Information emerging from DNA is expressed in the cytoplasm which in return imposes its action on DNA.

[1] The notion that information propagates from nucleus to cytoplasm was clearly described in 1900 [*Garnier*, 1900]. Microscopic examination of cells taken at different functional stages permitted *Garnier* to draw the following conclusions: a substance different from chromosomes and elaborated in the nucleolus migrates to the cytoplasm where it forms a string of beads with cytoplasmic granules, a step which precedes protein biosynthesis. Once proteins have been synthesized, the string of beads disappears and a new cycle starts.

The activation, replication and transcription of genes are essential processes in cell development and are closely connected with the activity of RNA polymerases. Knowledge of the properties of these latter enzymes should permit a better approach to the problem posed by the release of specific information. Access of RNA polymerase to DNA and the amount of synthesized RNA will depend on how effectively two strands of a segment of DNA are separated and for what period. All substances capable of separating or closing DNA strands interfere in the activation or inactivation of genes, i.e., in the expression of their messages. Studies of the fixation of hormones, carcinogens, antimitotics, proteins and other substances biologically active on DNA, followed by sequence analysis of these covered segments, might clarify the difficult problem of how a given functional unit on DNA is defined. It is well known that hormones (steroids) control the expression of genes and the appearance of specific proteins, suggesting that they recognize specific sites on DNA in hormone target tissues [*Palmiter and Schimke,* 1973]. Besides their direct action on DNA, these substances may also exhibit a more or less direct effect on the activity itself of RNA polymerases or other enzymes, thus interfering in the release of specific information.

'Mosaic' Messenger RNAs in Eukaryotes

In inferior eukaryotes, mRNAs are not released in groups (operon, regulation unit) as they are in prokaryotes [*Pontecorvo,* 1963; *Weinstock* et al., 1978]. Evidence was produced that most single copies of pre-mRNA are formed from repetitive and active sequences. Once selected, active sequences are ligated to give a functional mRNA. In sea urchin eggs, repetitive and message sequences are dispersed [*Constantini* et al., 1980]. After the appearance of the first RNA copy (pre-mRNA), intervening sequences disappear and active ones are ligated (fig. 6), and then a polyadenylate segment (poly A) binds to 'mosaic RNA' by a covalent linkage.

The mechanism by which active sequences are selected for 'mosaic mRNA' formation is far from clear. Complexity is even increased by the fact that most eukaryotic mRNAs contain a methylated adenine per 500–1,000 nucleotides [*Shatkin,* 1976] and a polyadenylate segment of 10–200 residues [*Darnell,* 1978] bound by a covalent linkage to the 3′-terminus nucleotide. Only mRNAs for histones do not appear to possess the polyadenylate segment [*Darnell,* 1978; *Browerman,* 1970; *Kaufmann* et al., 1977]. The fact that actinomycin D strongly inhibits poly A segment synthesis in the cell nucleus

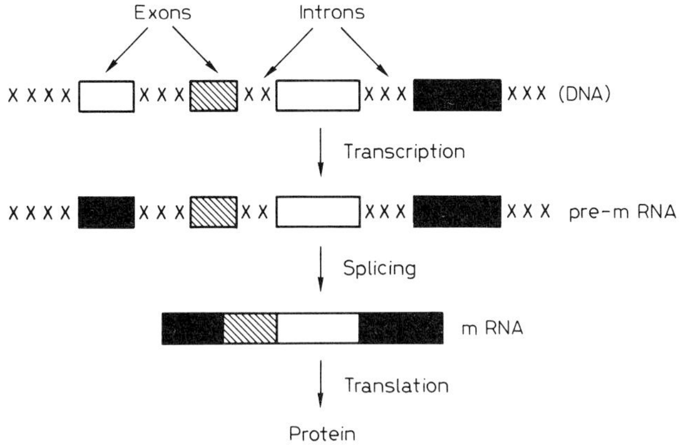

Fig. 6. Schematic representation of mosaic mRNA formation from pre-mRNA [*Gilbert*, 1978].

suggests that polyadenylate is a transcript of DNA [*Darnell*, 1978]. At present, scientists have to face the problem of the role of intervening sequences dispersed in pre-mRNAs. However, certain observations suggest that repeated sequences ahead of the beginning of the gene are required for transcription. One might wonder if some of these sequences are there to receive various stimuli: hormones, carcinogens or antibiotics, particularly when these substances are used in very low concentrations. It is known that several of them induce gene activation with the appearance of a particular mRNA. A well-defined portion of DNA, to be selectively transcribed needs to be defined and maintained as an 'activated' region of DNA. Basic studies, concerned with the fixation of hormones, carcinogens and other substances, should give more fundamental information for an understanding of the molecular process of gene activation and transcription into mRNA. We should recall that, in eukaryotes, formation of functional 'mosaic mRNA' from pre-mRNA involves the action of RNases and ligases. RNA fragments emerging from repetitive sequences might increase or decrease the transcription process.

Variety of RNA Polymerases

DNA-dependent RNA polymerases (EC 276) are complex enzymes containing subunits and are capable of transcribing DNA into complementary RNA. These enzymes link ribonucleotides (A, G, U, C) by the phosphodiester linkage, on a DNA template, after degradation of ribonucleoside-5′-triphosphates (ATP, GTP, UTP, CTP). In a gene which is a segment of double-stranded DNA carrying a given and stable character, only one of two DNA strands is transcribed into mRNA [*Hobom* et al., 1978; *Pieczenic* et al., 1975]. A gene may be defined as the unit of information assignable to an individual polypeptide chain. In eukaryotic and prokaryotic cells, several types of RNA polymerases are present. The function of each of these has not been completely elucidated. These enzymes catalyze initiation, elongation and termination of polyribonucleotide chains, a process called DNA transcription. RNA polymerase of prokaryotes [*Mirault and Scherrer,* 1972], probably translated from a polycistronic mRNA, contains at least four subunits: α, β, γ and σ [*Burgess,* 1971]. The eukaryotic nucleus contains three polymerases (I, II and III) whose existence was discovered by column chromatography of extracts from isolated nuclei [*Roeder,* 1975]. Their different functions have been classified according to their sensitivity to the inhibitory effects of α-amanitin, a toxic peptide [*Lindell* et al., 1970].

Nucleoli, isolated from nuclei, contain predominantly RNA polymerase I, the activity of which is not inhibited in vitro by α-amanitin [*Raynaud-Jammet* et al., 1972]. This peptide penetrates with difficulty into cells. RNA polymerase I appears to be responsible for the synthesis of precursors of ribosomal RNAs [*Roeder and Rutter,* 1970]. The nucleoplasm contains RNA polymerases II and III. The activity of enzyme II might be inhibited by low concentrations of α-amanitin: this enzyme appears to be responsible for the synthesis of heterogeneous nuclear RNAs (H nRNAs) in the nucleus. The activity of RNA polymerase III is inhibited by high concentrations of α-amanitin. This enzyme synthesizes 5S RNA and precursors of transfer RNA. RNA polymerases II and III appear to be involved in the transcription of DNA into mRNA and transfer RNA without transcribing other segments of DNA besides those we have cited [*Lindell* et al., 1978].

Although α-amanitin enables researchers to distinguish in vitro the activities of different polymerases, it was shown that, in vivo, transcription of ribosomal DNA (rDNA) into rRNA by RNA polymerase I was inhibited [*Lindell* et al., 1970], and these results contradict those obtained in vitro. When amino acids were removed from the culture medium, transcription in cells was greatly diminished. Addition of amino acids to the culture medium reestablishes a normal level of transcription in nucleoli [*Franze-Fernandez and Pogo,* 1971; *Grummt* et al., 1976]. However, it is known that amino acids as such are not required for the in vitro transcription of DNA into RNA. This may lead one to think that the in vivo synthesis of some particular proteins is needed in order to permit the process of transcription. In fact, some authors have shown that protein synthesis is a step which precedes the transcription of DNA into rRNA. Cycloheximide (an inhibitor of protein synthesis) provokes a rapid decrease of the activity of RNA polymerase I in the nucleus, while the activity of RNA polymerase II increases [*Lindell* et al., 1978]. It appears to us quite conceivable that proteins which interfere in the transcription of rDNA into rRNA may induce the separation of DNA strands, particularly in the region of genes responsible not only for the synthesis of rRNAs but also for that of mRNAs. Thus it was shown that anti-helix-destabilizing protein sera exhibit immunological cross-reactions with single-stranded DNA-binding proteins from various eukaryotic sources. It was possible to demonstrate that there is a striking localization of this protein in heat shock puffs (chromosomes of Drosophila salivary glands), known to be sites of new transcription [*Patel and Thompson,* 1980]. In contrast, RNA polymerases II and III, involved in the synthesis of mRNA, appear to be less dependent on this same protein.

Three RNA polymerases suffice to transcribe the thousands of genes necessary for the activity and differentiation of cells, but many other factors and molecules control and modulate the activity of these enzymes.

It is difficult to obtain homogeneous preparations of RNA polymerases from eukaryotic cells; these polymerases are enzymes with high molecular weight [*Seifart* et al., 1972; *Weinmann and Roeder,* 1974], all containing subunits. Even when purified, RNA polymerases are in most cases contaminated by RNase and DNase [*Chambon,* 1975] which makes it difficult to be certain of their respective in vitro functions and of factors playing in vivo a role in the liberation of specific information, cell differentiation in particular. Does each of these enzymes cause the transcription of certain given genes? This appears to be possible [*Hager* et al., 1976], although there is no proof along these lines.

Various Factors in the Activity of RNA Polymerases

RNA polymerases initiate the transcription of DNA into RNA, elongate and terminate the RNA chains and respect the complementarity of bases between DNA used as template and the newly synthesized RNA [*Chamberlin,* 1974; *Maitra and Hurwitz,* 1965]. These different steps cannot be accurately performed by RNA polymerases without the participation of other factors, namely molecules which are present in the neighborhood and the concentrations of which vary in an appreciable manner during cell metabolism.

In vitro, the activity of enzymes varies with different salt concentrations when saturating concentrations of DNA are used. RNA polymerase activity is much higher in the presence of Mn^{2+} than in the presence of Mg^{2+}; some basic proteins selectively stimulate the activity of RNA polymerase II when native DNAs are used as template [*Lentfer and Lezins,* 1972; *Lee and Dahmus,* 1973; *Seifart* et al., 1973; *Chuang and Chuang,* 1975; *Froechner and Bonner,* 1973], although the biological importance of these proteins has not been sufficiently analyzed. For example, we do not know if some of them play a role in the release of specific information. We should recall, for example, that basic proteins such as protamine easily bind to DNA or RNA, preventing nuclease action, and thus may interfere in the process involved in replication and transcription. It seems probable that some of the proteins which bind to the DNA double helix could contribute in opening the strands of DNA, thus increasing the appearance of single-stranded regions required for the activity of RNA polymerases. This interpretation is strengthened by observations that in vivo the activity of RNA polymerases depends on the presence of certain proteins whose properties have not yet been defined [*Gross and Pogo,* 1970; *Grummt* et al., 1976]. Among other molecules, steroid or peptide hormones, carcinogens and a number of particular RNA fragments are all capable of binding to single- or double-stranded DNA. They may either separate or close DNA strands, thus permitting or preventing mRNA synthesis, as the case may be. Both RNA polymerases A and B from *Drosophila melanogaster* larvae transcribe single-stranded DNA more actively than the corresponding native DNA [*Greenleaf* et al., 1976; *Losick and Chamberlin,* 1976]. The failure of wheat germ RNA polymerase II to wind or unwind DNA served to demonstrate that DNA has to be relaxed and the specificity in binding implies help from some unknown factor(s) [*Dynan and Burgess,* 1981].

DNA Templates Used by RNA Polymerases

RNA polymerases I and II from mammalian cells use single-stranded segments occurring in double-stranded DNA to initiate transcription, while RNA polymerase III easily initiates transcription of double-stranded DNA [*Sklar and Roeder,* 1975]. In vitro, RNA polymerases transcribe three types of DNA with varying degrees of success: single-stranded, double-stranded, and synthetic polydeoxyribonucleotides, sometimes even polyribonucleotides. The polymerase first has to bind to the DNA template. This results in the formation of a binary complex [*Richardson,* 1969] called the 'initiation complex' [*Maitra and Hurwitz*, 1965; *Maitra* et al., 1967]. The length of DNA covered by a molecule of RNA polymerase (holoenzyme-containing subunit σ) varies from 20 to 500 base pairs [*Ptashme* et al., 1976; *Williams* et al., 1973; *Schimamito and Wu*, 1980].

The binding of RNA polymerase to double-stranded DNA in order to transcribe it leads to the local separation of DNA strands, thus creating conditions for the transcription of given segments of DNAs, since only one of the two strands is transcribed into RNA by RNA polymerase I or II [*Hobom* et al., 1978]. The inactive strand of DNA modified by mutation, for example, does not influence the active strand which normally selects bases for mRNA synthesis [*Pieczenic* et al., 1975]. If each strand of DNA were used as template for mRNA synthesis, each gene would produce two mRNAs with complementary coding sequences for two different proteins. Genetic evidence shows that each gene appears to control the synthesis of only one protein, and hybridization techniques between DNA and RNA show that only one strand of DNA is transcribed into RNA [*Pieczenic* et al., 1975]. Once mRNA has been released, two DNA strands might reassociate into the double helix. Nicking of double-stranded DNA by DNase leads to the appearance of single-stranded fragments of DNA, and then RNA polymerase synthesizes 400 to 600 times more RNA than with intact double-stranded DNA. On a single strand of DNA template the RNA polymerase synthesizes a product which is often much longer than the template itself. This is particularly clear when a short DNA chain containing segments of thymidylic acid is transcribed into long poly-r(A) chains [*Chamberlin and Berg*, 1964a, b; *August* et al., 1962]. Access of RNA polymerase to initiation sites of DNA is critically sensitive to ionic strength as well as to the presence of dinucleotides. These latter molecules compete with ribonucleoside-5′-triphosphate (GTP and ATP) for the initiation of the RNA chain. It is clear from this that selective initiation may be enforced by certain dinucleotides [*Davis and Hyman*, 1970].

It is noteworthy that, in certain circumstances, RNA polymerase may catalyze the synthesis of polyribonucleotides rich in A and G nucleotides [*Milanino and Chargaff*, 1973] in the absence of template DNA. This is a matter of great interest but is not yet properly understood.

Initiation of Transcription

RNA polymerase-holoenzyme (RNA polymerase containing subunit σ) binds to the 'promoter' (site for binding RNA polymerase) and opens

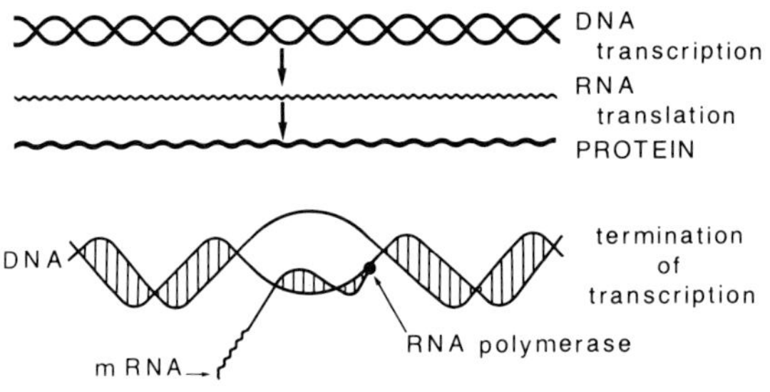

Fig. 7. Model of the coding strand DNA.

double-stranded DNA along 6–8 base pairs; the starting point for RNA synthesis is specified by the first purine ribonucleoside-5′-triphosphate (ATP or GTP) involved in this complex [*Chamberlin,* 1974, 1976]. There are two distinct sites on the enzyme, one which binds ATP or GTP only and another which binds the ribonucleotides to be linked up on the DNA chain, this site being considered as the site for chain elongation. Selection of nucleotides for chain elongation proceeds by the formation of hydrogen bonds with DNA used as template. When a phosphodiester bond is formed, the terminal nucleotide containing a 3′-OH group participates in the formation of this linkage with the next nucleotide (chain elongation). The enzyme is displaced by one base pair on DNA [*Hinkle and Chamberlin,* 1972; *Wu and Goldthwait,* 1969a]. As soon as the stop signal is given which announces the end of the transcription of a DNA segment, the enzyme is detached and binds on another binding site. Synthesized RNA is separated from DNA (fig. 7). During the initiation of transcription, RNase H (the enzyme which separates hybrid DNA-RNA) might, if necessary, remove the RNA primer which had been used to replicate DNA and then becomes useless. It was suggested that the 'active' S nRNA (small nuclear RNA containing 160–175 nucleotides) might recognize promoter or regulator sequences in the DNA, thus providing the extra accuracy which is essential to the release and maintenance of specific informations in eukaryotes [*Liu* et al., 1978].

The RNA transcripts, obtained in vivo by RNA polymerase, are molecules far smaller than the DNA used as template. For this reason it was supposed that RNA polymerase would bind to different initiation sites on DNA. DNA strand separation locally induced by RNA polymerase does not need the presence of nucleoside-5′-triphosphates, is not prevented by rifampicin, takes place at 37 °C (not at 0 °C) and is inhibited by high concentrations of KCl. In fact, at high doses, KCl prevents DNA strand separation altogether, thus making it impossible for RNA polymerase to bind to DNA and to separate the two strands [*Wang* et al., 1977].

The binding of certain substances to DNA, even close to the initiation site, might result in a reduction of efficiency for the DNA-RNA polymerase 'complex' formation (± nucleoside-5′-triphosphate, ATP or GTP). These results obtained in vitro with purified enzyme and DNA cannot be rigorously extrapolated to what is happening in vivo, in a biological environment.

Accuracy of DNA Transcription

Controversy surrounds the accuracy of in vitro transcription of DNA into RNA. According to some authors [*Zylber and Penman*, 1971; *Brown and Gurdon*, 1977], in vitro transcription of DNA or chromatin only completes the RNA chains which were already initiated in vivo, prior to isolation. Depending upon the moment of isolation, this material might contain RNA already synthesized which binds to DNA. Thus the promoter site is not free for RNA polymerase. RNA must first be removed from DNA, leaving the initiation sites free for fresh new RNA chain synthesis.

This does not always happen, and one faces the possibility of drawing incorrect conclusions about in vitro RNA synthesis with chromatin as template. Thus it was essential to use purified DNA, some of whose nucleotide sequences, corresponding to initiation sites, have been analyzed [*Gilbert*, 1976]. The promotion region for RNA synthesis is spread over 20–50 nucleotides [*Giacomoni* et al., 1974; *Walter* et al., 1967]. Most of our knowledge concerning the fixation of RNA polymerases to DNA comes from studies carried out in vitro to determine various steps, ranging from initiation through elongation to termination of RNA chains. These steps are followed by the release of synthesized RNA (fig. 7). Several components present in the extract from human KB cells are essential for accurate initiation of transcription by purified RNA polymerase II [*Matsui* et al., 1980]. However, the existing cell-free transcription systems fail to reproduce the in vivo pattern of gene regulation. Extensive extrapolation of these data to the in vivo process remains difficult. For example, *E. coli* DNA-dependent RNA polymerase or *Xenopus* RNA polymerase synthesizes in vitro heterogeneous RNA chains longer than 5S and originating from *Xenopus* erythrocytes when DNA is used as template [*Brown and Gurdon*, 1977]. This DNA also contains a 'spacer DNA' which is not an integral part of thc gene. In contrast, when 5S DNA is injected into a *Xenopus* oocyte, it is correctly transcribed into RNA, starting at the beginning of the corresponding gene, without transcribing a 'spacer DNA' of the 5S DNA preparation. This example shows that in vivo transcription of a particular gene is accurately carried out while this is not always the case in vitro. On the other hand, in vivo, cytoplasmic molecules (initiation and termination factors) must interfere to define the segment of DNA to be copied [*Brody* et al., 1970].

Initiation of Transcription and Rifampicin

Rifampicin has been particularly used for studies of the in vitro transcription mechanism using bacterial DNA-dependent RNA polymerase [*Lill* et al., 1970; *Krakow* et al., 1976; *Chamberlin,* 1974]. This antibiotic inhibits initiation but not elongation of RNA chains when the reaction has started in the absence of the antibiotic [*Maitra and Hurwitz,* 1965], which binds strongly to *E. coli* RNA polymerase. It can be removed from the enzyme in the presence of sodium dodecyl sulfate or high concentrations of guanidine-HCl. Once released, the enzyme possesses full activity [*Neuhoff* et al., 1970]. Rifampicin appears to be a highly specific inhibitor for bacterial RNA polymerases and has no effect on corresponding enzymes from mammalian cells. Rifampicin does not inhibit the formation of the first phosphodiester linkage which is catalyzed by the enzyme linked to DNA [*Johnson and McClure,* 1975; *Wu and Goldthwait,* 1969b; *Rhodes and Chamberlin,* 1975]. Preincubation of *E. coli* RNA polymerase with a template DNA and purine nucleotides (ATP or GTP) leads to the formation of the 'enzyme-DNA complex'. If rifampicin is added before the addition of nucleotides, dissociation of the complex is possible in the presence of high salt concentrations [*Anthony* et al., 1966; *Stead and Jones,* 1967; *Di Mauro* et al., 1969]. Thus, resistance of the 'enzyme-DNA complex' to dissociation is probably due to the presence of a short RNA chain synthesized before the addition of rifampicin [*Khesin* et al., 1967]. The function of GTP in the binding of RNA polymerase to promoter sites on DNA [*Bautz and Bautz,* 1970] and resistance of RNA polymerase to rifampicin in the binary complex with DNA were reported [*Sippel and Hartmann,* 1970].

Initiation sites for RNA polymerase have been sequenced for several DNAs after isolation of a DNA segment protected by RNA polymerase from the DNase effect [*Schaller* et al., 1975]. Nucleotide sequences and the size of the DNA segment corresponding to the promoter were evaluated at 41–44 base pairs. If the isolated promoter DNA segment is used as template, RNA polymerase synthesizes polymers containing 17–20 bases corresponding to the 5′ end sequence of RNA.

It will be of interest to examine the locus for mutation which blocks or stimulates transcription [*Gilbert,* 1976]. In addition to essential sites for fixation of RNA polymerase to DNA, there exist minor promoter sites. When the σ subunit is lost, the enzyme broadly synthesizes RNA without any selection of different DNA segments [*Burgess,* 1971]. In general, RNA polymerase binds wherever double strands are locally separated, and any further enzyme destabilizes hydrogen linkages, leading to enlarged strand separation. It should be noted that RNA polymerase has the capacity to recognize specific sequences on the external surface of a double-stranded DNA, possibly more easily accessible to the enzyme.

RNA Chain Elongation

The fundamental mechanism for RNA synthesis is very similar to that used for DNA synthesis. Polymerization of nucleotides proceeds in a given direction. Analysis of synthesized RNA chains on template DNA has shown that the newly attached nucleotide in the RNA chain is a terminal nucleotide with the 3′-OH free group. The nucleotide found where synthesis starts contains, in the 5′ position, a triphosphate group, indicating that RNA synthesis proceeds from 5′ to 3′ [*Watson,* 1965]. In addition to bivalent ions, all four ribonucleoside-5′-triphosphates are strictly necessary to synthesize a complementary RNA. Omission of only one of the four nucleotides considerably diminishes the synthesis and stops the formation of complementary RNA. Stabilization of RNA chains during synthesis appears to be partly due to RNA polymerase which has lost its σ subunit. It has been suggested that this subunit of RNA polymerase might play a role in the prolonged expression of an operon, i.e., several transcription cycles through numerous cellular divisions [*Errington* et al., 1974]. There exist endogenous and exogenous molecules capable of inhibiting elongation of RNA chains in the presence of RNA polymerases, as was shown for inhibitors of the initiation step. Consequently, the activity of RNA polymerase II is inhibited by low concentrations of α-amanitin (10^{-10} *M;* 10^{-7}*M*) [*Lindell* et al., 1970]. This toxic peptide inhibits the elongation but has no effect on the fixation of nucleoside-5′-triphosphates to DNA. In the presence of α-amanitin there is no dissociation of RNA or DNA from the ternary complex ‘enzyme-DNA-RNA’. Inhibition of RNA chain elongation can be observed during synthesis of RNA. Actinomycin D inhibits RNA chain elongation but not initiation: this antibiotic differentially inhibits the synthesis of various types of RNA, rRNA in particular [*Lindell*, 1976]. Addition of α-amanitin at high concentration (400 μg/ml) to a medium containing isolated nuclei inhibits 4S and 5S precursor RNA synthesis, while RNA polymerase III, under these same conditions, is inactive [*Weinman and Roeder*, 1974]. The α-amanitin does not freely penetrate into eukaryotic intact cells, and thus the results, concerning selective inhibition of specific RNA polymerases in vitro and in vivo with α-amanitin, are scarcely interpretable. In contrast, results obtained with nuclei treated with this toxin are consistent with results obtained using purified enzymes (same sensitivity to α-amanitin).

In order to get an accelerated transcription corresponding to a large increase in mRNA, the synthesizing system must either have more single-stranded DNA templates at its disposal for a given amount of enzyme or else must function over a longer period of time. In order to be transcribed after separating the strands, DNA must be maintained in such a way as to be accessible to RNA polymerase but not to DNA polymerase, i.e., priority must be given to transcription over replication. These conditions are favored by the helical structure of DNA and by cytoplasmic factors, whose nature has to be defined. Maintenance of two strands of a DNA segment in a separated position is one of the essential steps needed for the release of specific information. So far, we do not know enough about it.

RNA Chain Termination

Transcription of a portion of DNA corresponding to a genetic unit (gene, operon, cistron) leads to the synthesis of complementary mRNA of the size of the corresponding transcribed genetic unit. After initiation and chain elongation, signals are emitted to terminate the process. What are the signals and biological phenomena at the origin of this termination? Various hypotheses and experimental results have been suggested as an answer to this question. Data obtained in vitro during transcription of DNA into RNA as well as genetic results and considerations suggest that signals for the termination of the RNA chain are located at the level of terminal regions of mRNA containing 6–8 uridylic acid residues. The regions of RNA chains containing segments rich in U nucleotides might facilitate dissociation of RNA from DNA. The enzyme which appeared to 'slide' on a DNA strand will stop to function. It was suggested [*Gilbert*, 1976] that sequences of mRNA rich in G-C nucleotides which immediately precede U nucleotides might contribute to an unambiguous signal permitting RNA polymerase to transcribe a given length of DNA. Recently, results have been reported providing an explanation for the finding that most primary RNA transcripts terminate with five or more consecutive rU residues, but not rA residues. This finding suggests that instability of the DNA-RNA hybrid at the growing point of transcription plays a role in the termination of transcription [*Martin and Tinoco*, 1980]. The cytoplasmic environment (molecules and factors determining the rhythm of replication of DNA and responsible for desynchronization of mitoses) must certainly participate in transcription and emit the stop signal to terminate the RNA chain. It is conceivable that small-size RNA or peptides or proteins binding to DNA could be an important factor in RNA chain termination, although decisive proof has yet to be produced.

Transcription of DNA in Chromatin

The extraordinary complexity of chromosomes makes investigation to determine the molecular process which operates in the transcription of DNA a delicate matter. It is a trap because the isolation of nucleosomes from the cell nuclei knocks out all processes of synthesis. Thus, in vitro, one can simply observe the continuation of the RNA chain synthesis, already started in the cell from which chromatin has been isolated. In addition, this synthesis must be carried out under conditions closely approximating those present in a cell in order to get a complete RNA chain synthesis. Judging RNA synthesis by hybridization of in vitro synthesized labelled RNA might give a false impression of RNA synthesis when in reality one is dealing with RNA chains simply completed in vitro.

The necessity to 'activate' chromatin has been suggested to explain the transcription of DNA into RNA [*Allfrey*, 1971; *Weintraub* et al., 1976]. The word 'activation' is not as well defined as it should be. However, this expression implies first of all reduced interaction between DNA and histones which, in chromatin, modulates or modifies the conformational state of the nucleosome, thus permitting the opening of nucleosome cores and access to RNA polymerases [*Weintraub* et al., 1976]. Activation of chromatin can be performed in vitro by salts which contribute to the transcription of DNA into RNA although they maintain the essential structure of chromatin. One can also destabilize the nucleosome by submitting chromatin to the action of endonuclease ECOR I, but this leads to the destruction of the structure of chromatin [*Wasylyk and Chambon,* 1980]. Modification of the level of specific sites of the primary structure of histones may be achieved by acetylation or phosphorylation and might be connected either with the activation process [*Allfrey,* 1971] or with the assembly of chromatin [*Ruiz-Carillo* et al., 1975]. However, nuclei from HeLa cells cultivated in the presence or absence of Na-butyrate, an agent which suppresses the deacetylation of histones, do not show any difference regarding in vitro transcription by either endogenous or exogenous RNA polymerase II. From these data it appears that the hyper-acetylation of histones does not modify the transcription of chromatin [*Lilley and Berendt,* 1979]; histones are an obstacle to RNA polymerase [*Philipson* et al., 1971], decreasing the length of synthesized RNA by 400–600 nucleotides, which corresponds to the size of a cistron, i.e., a functional unit. Recognition sites for RNA polymerase are located on regions of DNA which are not covered by the histones. It was suggested that acetylation of histones could be necessary for gene derepression [*Sarkander* et al., 1973; *Sarkander and Knoll-Köhler,* 1978]. If this were so, acetylation would thus increase the number of sites for RNA polymerase. This was detected in an in vitro system [*Oberhauser* et al., 1978]. In fact, acetylation of chromatin isolated from rat liver increases the number of initiation sites for RNA polymerase twofold. However, in this connection, contradictory results were reported, thus making a definitive conclusion difficult, as far as the importance of the acetylation process in the transcription of chromatin DNA is concerned.

Small Nuclear RNA and Transcription of DNA

Small RNAs have been isolated from the nuclei of different cell species [*Benjamin* et al., 1966; *Shih and Bonner,* 1969; *Britten and Davidson,* 1969; *Weinberg,* 1973]. The role of these RNAs containing 65–200 nucleotides in gene activation has been postulated [*Zieve and Penman,* 1976] but not proven. Small RNAs are present in the nucleus during the interphase. They migrate into the cytoplasm during cell division and are found almost intact in the nuclei of cells resulting from the cell division process [*Rein,* 1971; *Goldstein,* 1974]. We suppose that they may initiate and stop translation of mRNA in the polysomal system. During anaphase, nucleolar and nuclear RNAs are largely associated with chromosomes [*Goldstein,* 1976]. One of them (U_3-RNA) is localized in the nucleus. It was suggested that its role consists in stabilizing complexes with genes [*Busch,* 1976].

In vitro, small RNAs appear to stimulate the replication of chromatin DNA when this material is used as template [*Kanehisa* et al., 1972, 1974]. It was suggested that in cells which will undergo mitosis, these RNAs might be involved in selecting genes for transcription. Their stability and size are important [*Beljanski* et al., 1978c; *Blinkerd and Toliver,* 1974].

Nuclear RNA may have a positive role in gene expression, probably by a mechanism involving base pairing which might lead to local DNA strand opening. Such a model suggests that the amount of transcribed RNA must be high immediately after mitosis. Small nuclear RNAs were shown to stimulate transcription of DNA into RNA in a relatively specific manner [*Kanehisa* et al., 1974]. According to this model, acid nuclear proteins would remove nuclear RNA from their loci on the chromosome, which should then result in halting transcription [*Dienstman and Holtzer,* 1975].

Peptides and DNA Transcription

We have seen that different proteins (unwinding, untwisting proteins, gyrases, ATPases etc.), by binding to double-stranded DNA, induce DNA strand separation in order to liberate local regions of single-stranded DNA serving as template for DNA polymerase as well as for RNA polymerase. The proteins do not prevent the in vitro activity of these two polymerases. The importance of basic peptides such as neutropsin or distamycin A is illustrated by observations that they bind to double-, rather than single-stranded DNA. Interacting with the minor groove in DNA, they form a bridge between DNA strands. This interaction with DNA prevents RNA polymerase from using DNA as template [*Zimmer* et al., 1971]. Hydrogen bonds are formed between basic groups of these peptides and oxygen of phosphates of A-T pairs with a preference for G-C pairs. The linkage between oligopeptides and DNA is very strong: in order to destroy this complex high salt concentrations or organic solvents are required [*Krey* et al., 1973; *Larson and Wells,* 1974]. The increased hyperchromicity under the effect of basic peptides (unwinding of double-helical structure) indicates their possibly important role in replication and transcription. The origin, accumulation and disappearance of basic peptides should be studied in detail since their role in transcription might be essential. Do they appear when proteases are activated by certain cations? Acid copolypeptides (Poly-glu^1, Tyr^1 or Poly-glu^3, Phe^1) are powerful inhibitors of DNA-dependent RNA polymerase, while glutamic or aspartic acid polymers are inactive in this respect. The peptide chain containing the aromatic group Glu^1, Tyr^1, is a most powerful inhibitor. Inhibition is prevented by the addition of an excess of enzyme, suggesting that these peptides interact with RNA polymerase itself [*Krakow and Van der Helm,* 1970]. It should be recalled that this enzyme contains two molecules of Zn, strongly bound, which makes RNA polymerase susceptible to the binding of acidic peptides or other compounds [*Scrutton* et al., 1971]. Thus, some acid drugs (Congo red, Gallin, etc.) interact with the Zn present in the enzyme as well as with positively charged parts of the enzyme, including the site required for binding of template. The effect of various peptides acting on DNA-dependent RNA polymerase might be connected with the activity of proteases during cell division. In fact, it has been shown that neutral proteases such as purified trypsin can act in vivo as mitogens for some cell types [*Brown and Kiehn,* 1977; *Grumm and Armstrong,* 1979]. On the other hand, in mitogen-stimulated leukocytes, DNA synthesis could be suppressed by the product obtained by trypsin digestion of protein A isolated from *Staphylococcus aureus* [*Cowan* et al., 1979]. All these data could be of more fundamental importance if the role of peptides could be firmly demonstrated in the release of specific information.

Plant Hormones and DNA Transcription

Auxin (indolyl-3-acetic acid = IAA), a plant hormone used in small concentrations, initiates DNA synthesis in plant cells without inducing mitosis as a consequence of DNA-increased synthesis. The appearance of mitosis requires either kinetin (plant hormone) or high concentrations of auxin. Furthermore, other results indicate that IAA interferes in the synthesis of RNAs. Does this hormone act directly on chromatin or on DNA itself? It was shown that while low auxin concentrations inhibit dissociation of the chromatin structure, high concentrations induce it [*Fellenberg and Schöemer,* 1975]. Different possibilities for auxin action in RNA synthesis have been discussed [*Leshem,* 1973]. According to one of them, auxin might be involved in gene expression. Thus, for example, long-term treatment of lentil roots with auxin (2.5×10^{-4} *M*) leads to increased activity of RNA polymerase 1b [*Penon* et al., 1975]. One of the enzyme factors (γ) required for the initiation of transcription undergoes modification in the presence of auxin. The late effect of auxin on rRNA synthesis is connected with an increased level of nucleolar RNA polymerase [*Teissere* et al., 1973; *Guilfoyle and Hanson,* 1974]. Phosphate, an inhibitor of RNA chain elongation, is without effect on the γ factor whose activity may be suppressed by protease. Its synthesis is inhibited by cycloheximide. The γ and σ factors act on a level of recognition of initiation regions on DNA wherever RNA polymerase starts transcription by opening DNA strands. The role of γ and σ factors in DNA strand separation was not yet studied.

The amount of RNA polymerase increases in the presence of IAA [*Trewavas*, 1968], although the mechanism was not clarified. According to some authors, auxin appears to control the amount of a repressor factor (ribonucleoprotein) involved in the expression of genes, which is a rather vague interpretation of the action of the auxin [*Leshem and Galston*, 1971]. Further studies have shown that when injected into the subapical section of *Pisum sativum*, auxin induces an increase in RNA synthesis. This was first detected in the nuclear fraction of the cell and thereafter in the cytoplasm. It is of importance that newly synthesized RNA molecules in the presence of auxin have their GMP/AMP ratio increased [*Trewavas*, 1968]. Small changes in the base composition of mRNA were also noted. These facts may be of great interest if one keeps in mind that oncogenic RNAs or RNAs active in cell differentiation as well as transforming RNAs contain an increased amount of A and G nucleotides [*Beljanski* et al., 1971, a 1972 a–c].

The majority of RNAs synthesized in the presence of auxin are rRNAs, which suggests that the main action of the hormone is exhibited at a level of genes responsible for the synthesis of rRNAs.

Several important facts should be kept in mind:

(1) auxin, an unstable molecule, may temporarly inhibit the activity of RNases;

(2) RNases degrade rRNA, giving rise to RNA fragments, capable – in the presence of auxin – of inducing plant tumors [*Beljanski and Aaron-da Cunha*, 1976; *Le Goff* et al., 1976];

(3) the increased amount and the activity of RNase during tumor induction are maintained on a much higher level in plant tumor cells than in normal cells [*Reddi*, 1966; *Le Goff and Beljanski*, 1981];

(4) the amount of extractible ribosomal 23S RNA from oncogenic bacteria (tumor-inducible bacteria of crown-gall tumors) is negligible compared to the amount of this same RNA from bacteria which have undergone various treatments [*Le Goff*, 1971]. Thus, 23S rRNA is extracted from oncogenic bacteria with difficulty because it is rapidly degraded by RNase. It appears that the amount of extractible 23S RNA is connected with the oncogenic power of this bacterium [*Beljanski* et al., 1972a];

(5) auxin, at given concentrations, temporarily stimulates the synthesis of rRNAs in stem segments of plants [*Trewavas,* 1968]. There appears to be a correlation between degradation by RNase of ribosomal RNA and plant development.

In chapter 6, we shall describe our observations on the synergic action of IAA and small purine-rich RNA in the induction of tumors in plants. On the other hand, we have shown that necrosis of crown-gall tumor cells caused by the presence of particular RNA fragments differing from oncogenic RNA [*Beljanski* et al., 1978b], can be effectively antagonized by IAA [*Le Goff and Beljanski,* 1981]. In fact, we have demonstrated that auxin alone induces a low degree of hyperchromicity (8%) of DNA from healthy plants and a high level in DNA from either *Agrobacterium tumefaciens* B_6 (oncogenic strain) (15%) or from crown-gall tumors (20%), which tumors were induced by the same bacteria. DNA strand separation is dependent on IAA concentration. Under the same in vitro conditions, auxin does not appear to separate the strands of DNA originating from healthy and cancerous cells of mammalians.

Concordant evidence shows that auxin may act directly on DNA, modifying the threshold of DNA strand separation, resulting in the increase of DNA synthesis. Hormone instability, concentration and its effect on RNase create conditions which modulate replication of DNA through the RNA primer effect, without which DNA-dependent DNA polymerase cannot initiate replication [*Le Goff and Beljanski,* 1981].

During the germination of 'nung bean' seeds, one may observe a 7- to 10-fold increase of RNase; RNase is synthesized 24 h after the hydration of seeds, while protease synthesis starts 3 days later [*Chappell* et al., 1980; *Baker* et al., 1974; *Baumgartner and Harris,* 1976; *Baumgartner and Matile,* 1976]. This suggests that mRNA for RNase preexists in seeds and is translated into enzyme 24 h after hydration. The rapidity of enzyme appearance indicates that this RNase might well be responsible for the production of the oligoribonucleotides required for replication of DNA and possible for the transcription and translation of mRNA into proteins.

Inhibition of Transcription by DNA-Binding Molecules

There are inhibitors of DNA-dependent RNA polymerases which, by binding to template DNA, prevent enzymes from transcribing DNA into RNA. This inhibition often occurs without any structural relationship between the two. We should cite actinomycins [*Reich and Goldberg,* 1970; *Smith,* 1965], acridine dyes [*Richardson,* 1966; *Nakata and Hurwitz,* 1967], peptides [*Zimmer,* 1975], anthracyclins [*Goldberg and Friedman,* 1971], etc. Actinomycins strongly inhibit the activity of RNA polymerase during DNA transcription by this enzyme [*Goldberg* et al., 1963]. The inhibition may be reverted by addition of an excess of DNA to the reaction mixture. Actinomycin D, the most frequently used inhibitor of RNA polymerase, forms a stable complex without covalent linkage with DNA. Mg^{2+} increases the amount of actinomycin that binds to DNA. It is known that Mg^{2+} binds to phosphates of DNA, the consequence of which is to bring together two DNA strands into a double helical structure. In order to observe the formation of an 'actinomycin-DNA' complex, it is necessary for DNA to be double-stranded and to have guanine in nucleotide sequences. Actinomycin D does not bind to synthetic deoxypolymers free from guanine, such as poly d(A-T) or poly d(T-C). The mechanism by which anthracyclins, such as daunorubicin, adriamycin or nogalomycin, interact with DNA, inhibiting transcription into RNA, is not clearly understood and depends on the concentrations of these substances. In fact, they intercalate between ring systems of pair bases in the DNA double helix [*Goldberg and Friedman,* 1971] but when used at very low concentrations one can observe DNA strand separation, particularly in the case of DNA originating from cancer tissues [*Beljanski* et al., 1981a]. It should be recalled that adriamycin or daunorubicin induce DNA strand separation in the nuclei of mammalian cells [*Center,* 1979]. In other different systems these anthracyclins permit the release of specific information. Their concentrations are a decisive factor in these phenomena.

Polyribonucleotides such as poly(U), poly-fluoro(U) or transfer RNA act as inhibitors of DNA-dependent RNA polymerase. This could be explained by partial electrostatic interaction with positively charged RNA polymerase and phosphates of template DNA. Fixation of these polymers to DNA results in the departure of the subunit σ [*Krakow and Ochoa,* 1963].

Summary

The very great complexity of interactions between various molecules, a number of RNA polymerases, the effects of different factors on their activity, their capacity to accurately transcribe DNA, the choice of template, all of these considerations make the interpretation and determination of the exact role of each element taken separately an extremely delicate task.

Studies of three essential steps in the transcription, initiation, elongation and termination of RNA chains also bring up several questions, some of which have been solved by using specific inhibitors, small biological molecules inhibiting or stimulating this or that function in order to determine the role of each. However, the activation of genes, i.e., conformational modulation of the DNA template, remains the crucial point in the release of specific information. RNA polymerase itself slightly induces the separation of DNA strands, but its narrow action cannot account for the sudden and definitely acquired changes in cells during the release of specific informations, particularly during cell differentiation. The cooperation of endogenous and exogenous substances is needed for this process which requires modifications of chromatin DNA. These modifications in DNA lead to the freeing of single-stranded DNA segments accessible to the enzyme for transcription. Since it has an enlarged and new surface at its disposal, RNA polymerase may function more efficiently on new accessible segments. During this step, other factors such as small peptides, hormones, small RNA, carcinogens and antimitotics participate in maintaining locally separated DNA chains, one of which is transcribed into RNA. The role of each of three RNA polymerases has not been elucidated as yet.

3. Hormones in the Release of Specific Information from DNA

Introduction

It is a well-known fact that the role of hormones in cell life is fundamental and comprises numerous aspects. For the subject we are dealing with we should concentrate our attention essentially on one class of hormones, the steroids, which may act at different cellular levels but whose ultimate target is DNA in the chromosomes. The essential function of steroid hormones is the regulation of normal biological processes, such as cell metabolism, differentiation and development. Their action on certain specific genes results in the release of specific information appearing as mRNA or as particular proteins. Although steroids are regulatory compounds in certain cell types (steroid targets), they can lead to a dysfunction in several organs (skin, uterus, prostate, breast), resulting in the appearance of cancerous cells [*Lacassagne*, 1932]. Estrogens and other sex hormones have long been known to be capable of stimulating cell division. Their frequent injection in high dosages to young mice during the first weeks after birth leads to an increase of cell divisions in various hormone target tissues and the appearance of cancerous cells in mammary glands. These observations are of crucial importance, since for the first time it was shown that steroid hormones, i.e., nonmutagenic compounds, may overactivate some genes by persistent influence, thus forcing them to escape the normal regulatory mechanism. In fact, the transformation of cells by steroid hormones, as well as by certain carcinogens, does not happen instantaneously. It always requires a persistent action of such substances on healthy cells over a period of time. The interaction of steroid hormones and nucleic acids has been studied in detail to determine the mechanism of their action, but before going into this subject, we should summarize very briefly the present state of knowledge about the organization of chromatin structure in eukaryotes.

Chromatin Structure in Eukaryotes

Specification of the mechanism by which endogenous or exogenous steroids influence DNA to release information requires some knowledge of chromosomes as carriers of genetic material.

Three levels of DNA organization exist in eukaryotic chromosomes [*Laemmli,* 1979]: the first is the fundamental structure of the DNA string, carrying nucleosome cores and containing histones; the second, the assembly of nucleosome chains into chromatin; the third, the winding of chromatin fibre in order to form compact chromosomes.

Recent data have provided evidence that the chromatin fibre visible under the electron microscope appears like a coiled particle with a diameter of approximately 100 Å. The chromatin particles are located on DNA strands as beads on a string [*Olins and Olins,* 1974; *Ris,* 1975; *Bonner* et al., 1968]. The DNA in between these particles is termed 'spacer DNA' (0–100 bases) and can be degraded by nuclease (DNase from *Staphylococcus*) more easily than the protected segment of DNA contained in the chromatin particles. Nucleosomes (fig. 8) contain two H_2A and H_2B histones, both rich in lysine residues, and H_3 and H_4 histones rich in arginine residues [*Kornberg,* 1974b]. The histone octamer in the nucleosome core protects DNA against nuclease attack. The length of these protected segments is 140 base pairs [*Axel,* 1975; *Sollner-Webb and Felsenfeld,* 1975; *Lohr* et al., 1977; *Morris and Morris,* 1976]. The constant repeat length observed in somatic cells is not inherited chromosomally but is established via a common intracellular pool of diffusible substances [*Sperling* et al., 1980]. This nucleosomal organization has been found in all samples of chromatin examined [*Noll and Kornberg,* 1977; *Show* et al., 1976]. A fifth type of histone (H_1) is localized on the external surface of the nucleosome and does not appear to belong to the nucleosome core; it would bind rather to spacer DNA [*Noll and Kornberg,* 1977; *Whitlock and Simpson,* 1976]. The contribution of H_1 histone is not clear. All histones bind to DNA with ionic linkages: basic groups of histones react with acid phosphates of DNA. Isolated chromatin usually contains RNA polymerase which seems to be an integral part of the genetic material [*Akinrimisi* et al., 1965]. If DNA is covered by histones or other proteins and molecules, DNA polymerase cannot replicate DNA, and RNA polymerase will not be able to transcribe it. At a given moment and at least for a length of DNA corresponding to a genetic unit, it is necessary to make DNA accessible to RNA polymerase. In vitro, salts up to 0.45 *M*, incubated with chromatin, favor the synthesis of long RNA chains (2,000 nucleotides) [*Lohr* et al., 1977]. In vivo, however,

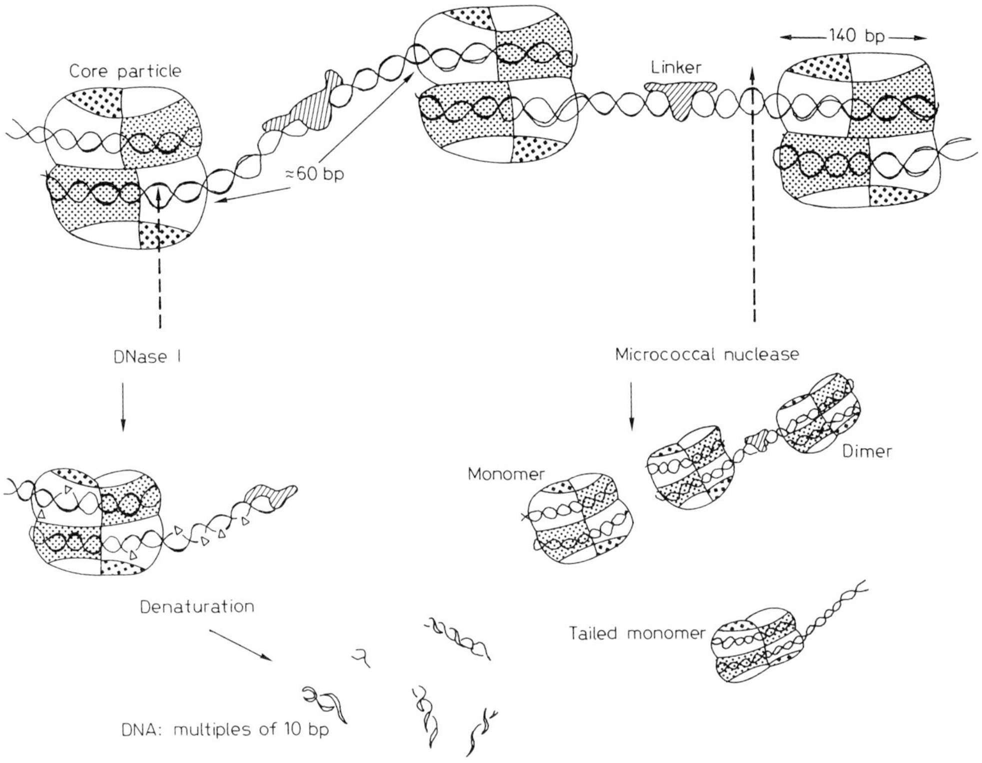

Fig. 8. Polynucleosomal structure of chromatin. Selective digestion by various nucleases. bp = Base pairs. [From *Smulson* et al., 1979, with permission.]

it would be extremely difficult to obtain such a high concentration of salts to activate the chromatin. Among different biological molecules involved in this process, steroid or other hormones play an important role in the release of specific information [*Dahmus and Bonner*, 1965; *Johnson* et al., 1978].

The dynamic behavior of chromosomes is complex and has, to a large extent, eluded the investigation of scientists. A great deal of research work has been devoted to determining the role and mechanism by which steroid hormones exhibit their action in the metabolism of nondividing cells or of cells which undergo differentiation, a process required for development [*Chandebois*, 1976]. When radioactive steroid hormones are injected into ani-

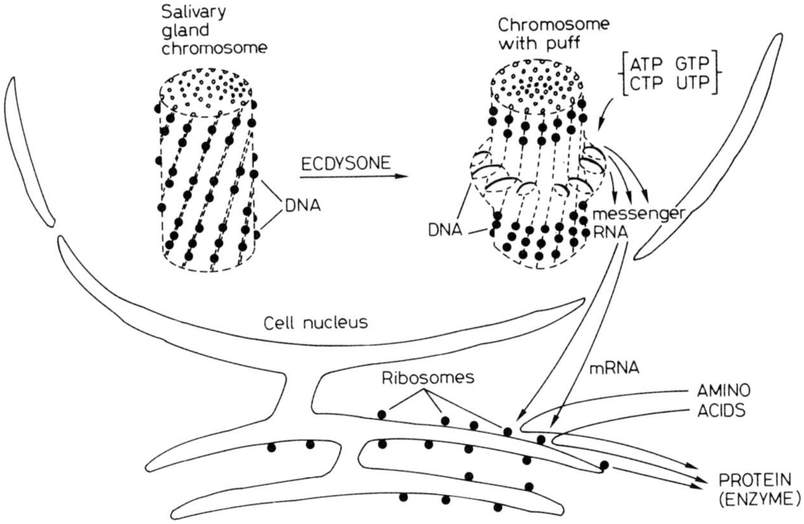

Fig. 9. Schematic representation of puffs on salivary gland chromosomes [from *Karlson*, 1963, with permission].

mals they concentrate essentially in the nuclei of the cells in hormone target tissues. Cell autoradiography reveals that testosterone and aldosterone concentrate on segments of DNA [*Loeb and Wilson*, 1965; *Edelman* et al., 1963]. Ecdysone (juvenile insect hormone) attaches to certain spots of DNA and induces some physiological changes there [*Clever*, 1963]. These are the 'puffs', where mRNA is abundantly synthesized [*Clever and Karlson*, 1960; *Pelling*, 1959] (fig. 9). As rRNA is synthesized in the nucleolus, ecdysone evidently activates specific mRNA synthesis rather than rRNA. The 'puffs' can be detected by fixation and chromosome staining followed by examination under a light microscope. Chromosomes with puffs have a destabilized structure. These observations are consistent with those showing that heterochromatin from lymphocytes, inactive in RNA synthesis, possesses a relatively dense structure [*Littau* et al., 1964] while euchromatin, active in RNA synthesis, is more relaxed [*Frenster* et al., 1963]. After the first experiments were completed, it appeared that steroid hormones favored the destabilization of helicoidal DNA strands, as was shown by heating DNA in solution [*Goldberg and Atchley*, 1966]. Thus, gene activity would be in direct relation with a given state of DNA, corresponding to its relaxed structure. We have shown the existence of a direct interaction of steroid hormones with a specific DNA,

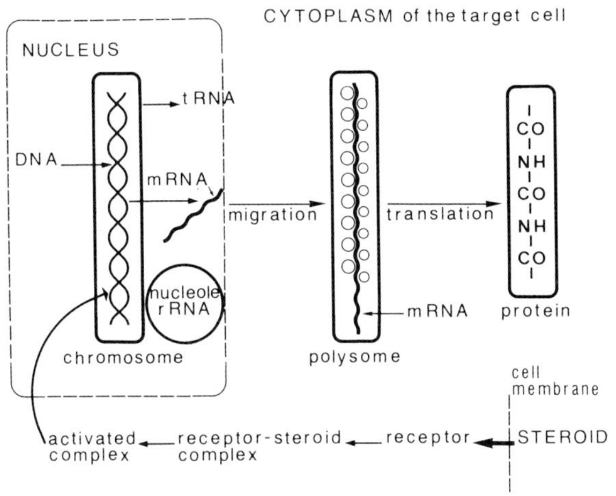

Fig. 10. Schematic representation of the mode of action of steroid receptors in the release of specific information from DNA.

resulting in an increase of hyperchromicity of the DNA and its template activity for DNA synthesis [*Beljanski,* 1979; *Beljanski* et al., 1981a]. The relaxed state of DNA is observed in the presence of steroid hormones when the DNA comes from steroid target tissues. Destabilization, already detectable with DNA from healthy tissues, is widely amplified (2- to 3-fold) with DNA isolated from cancer cells [*Beljanski* et al., 1981a]. However, according to the published research findings it seems that steroids do not act directly on DNA but rather through a protein receptor to which steroid hormones bind to form a 'steroid-receptor' complex. This confirms the fact that only DNAs isolated from steroid target tissues are destabilized by these hormones [*Clark and Gorski,* 1970; *Beljanski* et al., 1981a].

Steroid Receptors and Destabilization of Chromatin DNA

Interacting with a proper site on DNA in order to activate a given gene (or genes), steroid hormones force target cells to synthesize specific proteins [*Davidson,* 1968]. When an estrogen penetrates into the uterus or breast cells (fig. 10), it binds to the protein receptor present in the cytoplasm and then becomes activated. This activated complex, translocated into the nucleus, reacts with chromatin [*Gorski and Gannon,* 1976; *Liao,* 1975; *O'Malley* et al., 1976]. The hormone remains fixed in the nucleus for a certain time and then disappears [*Strumpf,* 1968; *Mohla* et al., 1972; *De Sombre* et al., 1972, 1975; *Clark* et al., 1977].

It is generally believed that the role of a receptor protein in the nucleus is to stimulate transcription, the hormone being necessary only for the translocation of the 'hormone-receptor' complex from the cytoplasm to the nucleus [*De Sombre* et al., 1972]. The hormone itself will not be involved in chromatin destabilization. Studies concerned with the mechanism of the attachment of the 'estradiol-receptor' complex either to chromatin isolated from nuclei or to purified DNA from hormone target tissues [*Shyamala and Gorski,* 1969; *Gannon* et al., 1976] reveal that DNA is indeed involved in this mechanism [*Shyamala-Harris,* 1971; *Clemens and Kleinsmith,* 1972]. It is conceivable that protein receptors for steroids once bound to chromatin may be degraded by proteases, thus providing peptides which could impose the length of DNA which has to be activated by hormones. The interference of certain poly- and oligoribonucleotides at low concentrations with a steroid-receptor complex, leading to the release of the receptor complex from DNA, was recently reported [*Liao* et al., 1980]. The receptor complex appears to have higher binding affinity toward single-standed deoxypolymers than to the double-stranded DNA. The activity of oligonucleotides that might bind to DNA or RNA depends on the length and on the type of nucleotides in the oligomers. Thus, poly(U-G) with uracil/guanine ratios of 1:5 were more active than poly(G) or poly(U). Active poly(U-G) oligo-ribonucleotides promoted the release of rat uterine estrogen and progesterone-receptor complexes and rat liver dexamethasone receptor complexes from DNA. A change in the local DNA helical structure, including partial strand separation, may occur during the binding and release of the receptor from DNA, creating binding sites with different affinities. The removal of RNA from DNA by the steroid-receptor complex may make the DNA template free for transcription. On the other hand, receptor binding to RNA may be involved in the post-transcriptional control [*Liao* et al., 1980]. Protein receptors for steroid hormones are apparently absent from tissues which are not target tissues [*Clever,* 1963]. Steroid receptors have been found for androgens in prostate cells [*Greenman* et al., 1965; *Liao and Tang,* 1969; *Magnan* et al., 1968] and even in the brain of newborn female rats [*Clayton* et al., 1970]. Protein receptors for glucocorticoids have been found in liver, lung, thymus, lymphoid cells, in fibroblasts and liver cells maintained in culture [*Clemens and Kleinsmith,* 1972; *Rousseau,* 1975a]. Receptors for androgens, glucocorticoids and progesterone have been found in most breast cancer tissues. Receptors for estrogens in these tissues are analogous to those detected in healthy tissues of hormone target organs [*De Sombre and Lyttle*, 1979]. Although it has been clearly established that steroids act through receptors as intermediaries in gene 'activation', it should be noted that hormones may also directly interact with DNA to induce separation of given single-stranded DNA segments, specific for steroids [*Beljanski* et al., 1981a]. This must result in the release of specific information which is detected in cells as specific mRNA or protein. The direct effect of hormones on DNA from hormone target tissues has been demonstrated by different authors [*Goldberg and Atchley*, 1966; *Beljanski* et al., 1981a; *Clemens and Kleinsmith*, 1972]. This effect is more pronounced with DNA originating from cancer tissues [breast, ovary; *Beljanski* et al., 1981a] than with DNA from corresponding healthy tissues. In vitro competition for the same site on DNA between a steroid hormone and an RNA primer involved in DNA replication of normal and cancer tissues has been observed [*Beljanski*, unpublished data]. If this is confirmed in vivo, which seems quite probable, then a locally arrested DNA replication would make room for the transcription of that particular DNA segment. We shall return to this problem later.

The fact remains that biologically activated DNA containing locally separated DNA strands permits the release of information to a greater extent than poorly activated DNA. This necessarily implies flexibility of the genome during cell differentiation with a major change of controlled regulation in genome activity.

Interaction of Steroid Hormones with DNA

The fact that steroid hormones are essentially in the nucleus of cells suggests that DNA must be the first site for fixation of the 'steroid-receptor' complex [*Yamamoto*, 1974]. Estrogen receptors appear to act by altering chromatin and, to some extent, polymerases – or both – in order to stimulate the synthesis of RNA [*Jensen* et al., 1974]. The fact that active chromatin is more relaxed and diffused than inactive chromatin [*Frenster*, 1965; *Cox* et al., 1973] led some authors to postulate that the biological activity of steroid hormones would consist in destabilizing the double-helical structure of DNA. However, no differential action of steroids has been observed in the presence of the steroid hormone between deoxyribonucleoproteins on the one hand and free DNA on the other [*Frenster*, 1965]. Thus, the problem remains unresolved.

However, various studies have shown that the fixation of steroids to DNA itself is possible. An increase in the temperature of a solution containing human placenta DNA and the addition of a steroid (1.3×10^{-5} M) causes increased DNA strand separation: absorbance at 260 nm of DNA increases more rapidly in the presence of hormone than in its absence. Thus 17β-testosterone, a biologically active hormone, accelerates DNA strand separation while 17α-testosterone, an inactive hormone, does not. Cortisol, insulin and *l*-epinephrine act in a similar way [*Goldberg and Atchley,* 1966] as well as 17β-testosterone.

By binding to the single-stranded segments of DNA which appear under the effect of heat, steroid hormones help to accelerate the process of strand separation. Guanine appears to be the target nucleotide for steroid action. The steroid effect obtained with human placental DNA is not observable with the DNA of *Bacillus subtilis* [*Goldberg and Atchley,* 1966]. It has been shown by equilibrium dialysis that steroid hormones do not bind to double-stranded DNA, only to denatured DNA or to a polyriboguanylate. In contrast, these hormones do not bind to poly(A), poly(C) or poly(U). In the course of this kind of research it was shown that steroid hormones interact with a short nucleotide sequence, a trinucleotide, for example. Thus progesterone forms a hydrogen bond between C-3 and C-20 (carboxyl group) and the 2′-amino group of guanine, if this latter occupies the positions 1 and 3 in the trinucleotide [*Cohen and Kidson,* 1969]. These observations suggest the destabilization of hydrogen linkages between DNA strands and explain the rapid separation of double strands of cancer DNA originating from steroid-hormone tissues [*Beljanski* et al., 1981a]. They would suggest that the steroid effect may result in accelerated DNA replication and also in the release of specific information which appears as specific RNA or protein.

Effect of Estrogens in DNA Replication

The DNA of eukaryotic cells contains replicative units lined up end to end, in the middle of which DNA synthesis proceeds in opposite directions, starting from replicative forks [*Huberman and Riggs,* 1968]. It has been shown by autoradiography that estrogen increases the lenght of the replicative forks of DNA [*Leroy* et al., 1975], thus shortening the duration of the S phase of the mitotic cycle in the uterine DNA of mice treated with steroid [*Galand* et al., 1967, 1971; *Das,* 1972]. Labelling DNA from uterine epithelial cells of mice with radioactive thymidine shows that the DNA of estradiol-treated animals contains labelled segments which are 50% longer than the DNA of the untreated mice. The S phase of mitosis is shorter, and the increase in the length of replicated strands is about 50%. The steroid accelerates DNA replication which necessarily implies the presence of single-stranded DNA regions in greater numbers. These zones are necessary for DNA polymerase activity. Used in excess, steroid hormones might act on the DNA of some epithelial cells, uterus, bladder or other tissues, stimulating DNA synthesis and, consequently, cell multiplication. In fact, a single dose of estrogen to young rats increases the activity of DNA polymerase α in the uterus 3- to 4-fold, and this results in an important increase in DNA synthesis [*Harris and Gorski,* 1978]. Multiplication of human breast cancer cells (hormone-dependent tissues) is accelerated by physiological doses of 17β-estradiol. This effect is due to the stimulated activity of DNA-dependent DNA polymerase [*Lippman* et al., 1976]. When the activity of DNA polymerase is inhibited in these cells by antiestrogen compounds, addition of 17β-estradiol restores this activity to a normal level [*Edward* et al., 1980]. It would be interesting to know to what degree steroids modulate the activity of RNases which cause cells to be richer or poorer in small RNAs or the oligoribonucleotides essential for DNA replication and mRNA translation (chapter 4). There is a competitive interaction between hormones and RNA primers in DNA replication. It should be recalled that in target tissues, cessation of estrogen administration is accompained by an increased synthesis of (2′,5′-)oligoadenylates which are inhibitors of protein synthesis [*Shimizu and Sokawa,* 1979]. Some of these proteins are involved in DNA replication. In the presence of interferon, an inhibitor of cell growth and division, an increase (10–100-fold) of (2′,5′)-oligoadenylate synthetase at the transcriptional level has been reported. Adenylic residues carrying the triphosphate end (2′,5′/pppApApA) activate RNase F, while (2′,5′)ApA competitively inhibits the 2′-phosphodiesterase. Changes in the RNase level and activities in the presence of oligoadenylates are probably responsible for the inhibition of DNA replication, transcription and, consequently, cell division [*Kimchi* et al., 1981; *Schmidt* et al., 1978]. Insulin or the epidermal growth factor [*Richeman* et al., 1976; *Henson* et al., 1966] stimulates DNA synthesis in hepatocytic cells cultured in vitro and originating from partially hepatectomized adult rat liver. After 3 days, the increase in the amount of DNA might be of the order of 35–50%, compared to 2–7% for the control DNA. It should be noted that DNA replication does not always correspond to cell division. DNA synthesis and cell division, induced by insulin in the explants of the mammary glands from pregnant mice, appear to be essential for the initiation of casein synthesis by prolactin and hydrocortisone [*Lockwood* et al., 1967]. Insulin has an effect on RNase activity [*Beljanski* et al., unpublished results] which may provide RNA fragments. These, in synergy with hormones, may contribute to DNA replication and release of specific information.

DNA isolated from hormone target organs (breast, ovaries), has its replication slightly stimulated by testosterone or estradiol. If the DNA comes from hormone-independent tissues, normal or cancerous, one does not observe any further stimulation by hormones. However, there is an exception to this rule. In vitro synthesis of neurocarcinoma DNA is stimulated by testosterone. It should be noted that hormone receptors have been found in neural tissues [*Strumpf,* 1968]. A stimulatory effect which is always much stronger for DNA of tumor cells than of normal ones may thus be observed [*Beljanski* et al., 1981a].

Phytohormones and DNA Synthesis

The plant growth hormone, auxin, stimulates the in vitro DNA synthesis of healthy plant cells and even more that of DNA from cancer cells. Auxin is one of the necessary factors for tumor induction in plants either by *Agrobacterium tumefaciens* or by particular purine-rich RNA [*Beljanski and Aaron-da Cunha,* 1976]. This phenomenon occurs through inactivation and/or activation of genes. We have demonstrated that slight or strong stimulation of DNA synthesis in vitro (depending on whether the hormone is in the presence of DNA from normal or cancer plant cells) corresponds to DNA strand separation, which is slight in normal and strong in cancer DNA. Direct evidence for gene activation under the hormone effect may be obtained in this way [*Le Goff and Beljanski,* 1981]. It has been reported that gibberellin A_7, one of the principal plant hormones, binds highly specifically to AT-rich DNAs [*Kessler and Snir,* 1969]. In combination with a DNA ligase, it induces the formation of circle-like loops in nuclear DNA [*Kessler,* 1971]. DNA isolated from in vitro cultured protocorms of the orchid *Cymbidium* [*Nagl and Rücker,* 1976] after treatment with various plant hormones shows the variation of the thermal denaturation profiles. Thus, in auxin-treated cultures, DNA contains AT-rich fractions while in gibberellin-treated ones the DNA shows the expansion of the G-C-rich fractions.

On the basis of these results it would appear that phytohormones control the differential replication of certain DNA sequences and, consequently, the transcription.

Actinomycin D and Steroid-Receptor Complex Fixation to Chromatin

We have seen that the steroid-receptor complex formed in the cytoplasm is transported to the nucleus of cells where it binds to segments of DNA more or less covered by chromatin. This effect was observed after progesterone administration in chicken oviduct [*Schadler* et al., 1972]. It has been further studied in various experiments in order to acquire a better understanding of the interaction between the steroid-receptor complex and chromatin. The amount of such a complex might be modulated by the presence of actinomycin D [*Horwitz and McGuire,* 1978], known to intercalate between DNA strands. Because of the chromophore which it contains, actinomycin binds to successive G-C base pairs, and pentapeptides are located in the minor groove of double-stranded DNA [*Goldberg and Friedman,* 1971; *Kleinman and Huang,* 1971]. Actinomycin appears to block the access to the estrogen-receptor complex which ultimately should bind to DNA. Inhibition of RNA synthesis observed in the presence of actinomycin D was not of the same magnitude for all RNA when small doses of the antibiotic were used. This suggests that, depending on the dose, actinomycin does not bind to the same binding sites on DNA, which is in agreement with our recent observations [*Beljanski* et al., 1981a]. It was noted that rRNA synthesis was particularly sensitive to actinomycin D. It was also shown that when one removes estrogen from a culture medium where a hormone has stimulated protein synthesis, this latter is arrested. If actinomycin D is added immediately after removal of the steroid, protein synthesis continues and proteins accumulate [*Horwitz and McGuire,* 1978]. In several laboratories, the appearance of specific proteins subsequent to estrogen treatment has been observed [*Schimke,* 1974; *Mayol and Thayer,* 1970; *Hilf* et al., 1965], and actinomycin causes a superinduction of some of these [*Tomkins* et al., 1972]. This protein superinduction by the antibiotic would be caused either through the initiation of protein turnover [*Thompson* et al., 1970] or the stimulation of mRNA translation [*Palmiter and Schimke,* 1973]. These last observations imply that actinomycin D might inhibit the activity of proteases and stimulate that of RNases. It is essential to keep in mind at this stage that hormones, as well as actinomycin D, depending on the dose used, inhibit or stimulate RNase activity [*El-Sewedy* et al., 1978; *Le Goff and Beljanski,* 1981]. On this will depend the activities of DNA and RNA polymerases through the intermediary action of oligoribonucleotides (chapters 1 and 2). Thus, the effect of actinomycin D in the superinduction of protein biosynthesis may be explained by its interference in the appearance of oligoribonucleotides, some of which stimulate the

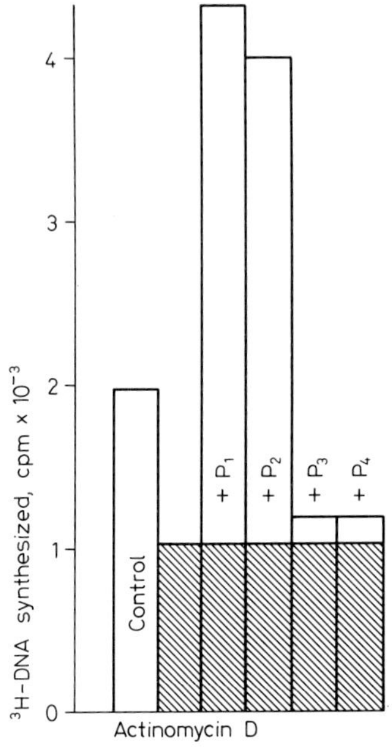

Fig. 11. Healthy lung DNA synthesis in vitro in the absence or presence of actinomycin D (2 μg/assay). Effects of RNA fragments P_1 and P_2 (rich in purines) or P_3 and P_4 (poor in purines) (2 μg/assay) [*Beljanski*, unpublished results].

initiation or translation. Also, superinduction of proteins induced by actinomycin D following cell stimulation by estrogen suggests that each of these substances must contribute to open DNA strands becoming more accessible to RNA polymerase. At the very least, a synergic action may thus be produced. The effects of actinomycin D and estradiol on some enzymes were studied in normal and tumorous mammary glands. Treatment with actinomycin plus estrogen led in both cases to the increase of the activity of several enzymes, among which glucose-6-phosphate dehydrogenase. This increase was not obtained by hormonal treatment alone [*Hilf* et al., 1965]. The enzyme increased activities have resulted from DNA-directed RNA synthesis. Here also, a synergist effect of actinomycin + estrogen is probably due to increased local DNA strand separation.

When 9,10-DMBA [dimethylbenz(a)anthracen] is added to the incubation medium, it stimulates the in vitro synthesis of breast cancer DNA:

estradiol or progesterone added after the carcinogen causes amplified DNA synthesis. In contrast, estrone addition slows down the DNA synthesis, which was stimulated by DMBA. These results should be compared with those showing that mammary cancer in mice induced by 7, 12-DMBA (a strong carcinogen) does not develop in ovariectomized animals [*Huggins and Grand*, 1966]. The addition of estrogens accelerates the multiplication of cancer cells. It appears that mammary cancer cell development requires the opening of DNA strands over certain limits dictated by 7, 12-DMBA in a DNA region specific for steroids. Such a synergy exists in in vitro conditions [*Beljanski* et al., 1981a].

A third example shows that the inhibitory action of actinomycin D, in normal or cancer DNA synthesis, might be suppressed efficiently by RNA fragments, used by DNA-dependent DNA polymerase I for DNA replication. For example, when lung cancer DNA synthesis is inhibited by actinomycin D, the addition of RNA fragments P_1, P_2 or P_3, variously rich in G and A nucleotides and each used separately, makes it possible to prevent inhibition [*Beljanski* et al., 1975, and unpublished results]. The RNA fragments P_1 and P_2 (very rich in A and G nucleotides) are less efficient than the RNA fragments P_3 and P_4 (less rich in G and A), under the same conditions. In fact, the latter completely suppress the inhibition produced by the antibiotic. In contrast, when actinomycin D inhibits the in vitro synthesis of DNA from healthy lung tissues, RNA fragments P_1 and P_2 not only remove the inhibition but also stimulate DNA synthesis, while RNA fragments P_3 and P_4 are practically inactive (fig. 11). Thus the inhibition caused by actinomycin D may be withdrawn by a certain amount of RNA fragments containing G and A nucleotides. The interference of steroid hormones in these events is of great importance.

Estradiol and Increase of RNA Polymerase Synthesis

The steroid-receptor complex controls RNA synthesis in chromatin. Its mechanism is not yet known, but the increase in RNA synthesis is detectable in the uterus 2 min after treatment of animals with estradiol [*Weckler and Gschwendt*, 1976]. 1 day after injection, the activity of nucleolar and nuclear RNA polymerases continues to increase in chicken liver and precedes that of proteins. This phenomenon was attested after enzyme purification. In contrast, if the activities of nucleolar and nuclear RNA polymerases increase, that of DNA polymerase is not modified, while RNase activity decreases by about 40%! It may be that this latter enzyme allows RNAs to accumulate over a certain period of time in order to splice them later on.

The total quantity of RNA polymerases I and III, extractable from the cells of adult rats which received estradiol 6 h before they were sacrificed, is not modified: both enzymes become activated only with time [*Weil* et al., 1977]. It was concluded that a 6-hour period is not sufficient to carry out chromatin modification. By way of contrast, 6 h after estrone or estradiol injection, an increased synthesis of RNA polymerases I and II in rat uterus can be observed. Estrogens interact not only with chromatin but also with DNA itself [*Beljanski* et al., 1981a]. Under the steroid effect, DNA locally undergoes an opening of double strands, and this allows the synthesis of mRNA and, consequently, of proteins. Injected concentrations of estrogens in the experiments cited above probably allow DNA strand separation in a relatively short time, which might explain the observation reported.

Steroid hormones appear to regulate the transcription of different classes of genes in mammalian tissues. In this respect, the activity of RNA polymerases in the uterus has been extensively studied. The pioneering work of *Gorski* [1964] has been followed by numerous other studies concerned with the increase of RNA polymerase I [*Weil* et al., 1977; *Roeder and Rutter*, 1970] and RNA polymerase II [*Glasser* et al., 1972; *Borthwick and Smelie*, 1975] under the hormone effect. Numerous authors agree in concluding that the activity of nucleolar RNA polymerase I increases after the first 6 h following estradiol injection [*Bieri-Bonniot* et al., 1977; *Glasser* et al., 1972]. In contrast, there is no such agreement in connection with the activity of RNA polymerase II [*Mohla* et al., 1972].

In fact, it was found that the activities of RNA polymerases I and II increase in a similar fashion during the first 6 h, a phenomenon not observed 24 h later. It was shown that in vivo, after 17β-estradiol administration, synthesis and increase of nucleolar and nuclear RNA polymerase activity precede a general increase of RNA and proteins [*De Sombre* et al., 1972]. The 17β-estradiol appears to induce in vivo the synthesis of a protein which, in vitro, stimulates the incorporation of ^{3}H-thymidine into DNA of 3T6 fibroblasts [*Schodell*, 1973]. Estrogens stimulate liver RNA synthesis in amphibians. This synthesis precedes that of proteins and is followed by the synthesis of RNA, apparently rich in U nucleotides, and then by an increase in ribosomal RNA synthesis. The RNA, rich in uridylate, was not characterized as 'DNA-like RNA'. Rat liver treated with cortisone first synthesizes and RNA rich in uridylate, then an RNA rich in guanine [*Wittliff* et al., 1972; *Yu and*

Feigelson, 1969]. When injected to adult ovariectomized rats, estrogens increase not only the rate of synthesis of uterine ribosomal RNA but also that of some other classes of RNA. These results have been interpreted as stressing the prevention of the wastage of high molecular weight RNA under the influence of estrogen, resulting in an increase of RNA [*Luck and Hamilton,* 1972]. Steroid action may contribute in modulating the activity of RNases. These enzymes are essential for the formation of mRNA and rRNA from precursor molecules, of RNA primers from rRNA, but are also needed for the destruction and elimination of useless RNAs. Some in vitro observations reported that inhibition of RNase activity takes place in the presence of steroid hormones [*El-Sewedy* et al., 1978]. In ovariectomized rats, the hormone causes an increase in RNAs, including rRNA [*Luck and Hamilton,* 1972], originating from high molecular weight precursor RNAs. The efficiency of this process necessitates the action of at least one specific nuclease. In fact, evidence for the existence of an endonuclease in the nucleolus has been reported [*Grummt* et al., 1979; *Perry,* 1976]: this enzyme cleaves precursor RNA on specific sites and liberates rRNA [*Mirault and Scherrer,* 1972]. This enzyme has been found associated with preribosomal particles and seems to be identical with the enzyme found in the nucleoli of HeLa cells. It possesses the properties of RNase III which leads to the appearance of rRNA [*Hall and Crouch,* 1977]. In addition, estradiol appears to exhibit its effect on RNA chain elongation during synthesis [*Barry and Gorski,* 1971], which can be interpreted as the consequence of the increased template DNA available for RNA polymerase [*Church and McCarthy,* 1970]. This seems to us the most probable effect especially in the light of recent results showing steroid hormone action on DNA strand separation [*Beljanski* et al., 1981a].

Auxin (IAA), a plant hormone, destabilizes the chromatin of healthy plant cells when used at high concentrations but is without effect at low concentrations [*Fellenberg and Schömer,* 1975]. Treatment of plant roots by IAA stimulates the activity of RNA polymerase [*Penon* et al., 1975]. The effect of auxin is manifest especially in rRNA synthesis, but for 3–4 h only [*Chappell* et al., 1980; *Barker* et al., 1974; *Baumgartner and Harris,* 1976; *Baumgartner and Matile,* 1976]. It may be imagined, and many experiments appear to indicate it, that after this period of time nucleases degrade RNA, thus furnishing RNA fragments capable of becoming primers for further synthesis. It should be mentioned that auxin acts directly on healthy plant DNA in order to separate DNA strands locally [*Le Goff and Beljanski,* 1981]. All the results of which we have spoken have found expression in a general stimulation of biological processes.

Steroid Hormones and Release of Information in Castrated Animals

It was first suggested and then reported by several authors that DNA template activity for RNA polymerase II is increased after treatment of animals by steroids [*Cox* et al., 1973; *Barker and Warren,* 1966; *Teng and Hamilton,* 1968]. 2 h after a testosterone injection to castrated rats, a quantity of highly varied nuclear and polysome RNA appears in nuclear RNAs. This suggests a certain post-transcriptional regulation by hormones. In vitro DNA-RNA hybridization indicates (if one assumes that 18,000 nucleotides are implicated in a message) that at least 69,000 messages are transcribed in the nucleus 2 h after treatment by testosterone, while 5,000 messages appear on polysomes after a stimulatory effect by an androgen. These results suggest that after testosterone administration the majority of RNAs appear in the nucleus where they have been formed and only a small quantity penetrates into the cytoplasm of the cell. For instance, the amount of ovalbumin mRNA is 13 times higher in the nucleus than that found bound to polysomes in the cytoplasm [*Hiremath and Wang,* 1979]. This is not surprising since steroid hormones, which cause the appearance of single strands in the DNA template, allow RNA polymerase to synthesize greater amounts of RNA than in the absence of these hormones.

Control of RNA and Protein Synthesis by Hormones

Numerous observations lead to the conclusion that increased RNA synthesis is the consequence of the administration of steroids or other hormones. Injection of labelled steroid hormones to animals or the incubation of these hormones with in vitro cultured cells is first manifest by the appearance of a hormone-receptor complex in the cytoplasm, which is later translocated to the nucleus [*Gennon* et al., 1976; *Shyamala-Harris,* 1971]. Under the effect of the hormone, heterogeneous RNAs appear in the nucleus. Nuclei isolated from animals pretreated with androgens synthesize much more RNA than nuclei from control animals [*Williams-Ashman* et al., 1964]. Identical results were obtained in mice kidney after treatment with androgens [*Kochakian,* 1969]. Under the influence of progesterone, new RNA species are synthesized [*O'Malley and McGuire,* 1969]. The synthesis of particular mRNA may be followed by the characterization of either RNA itself or of the corresponding protein synthesized in the cell or in an in vitro system. The synthetic glucocorticoid dexamethasone stimulates the synthesis of a small number of specific proteins, while 5-bromodeoxyuridine (BrdU) blocks the accumulation of secretory proteins [*Walther* et al., 1974; *Githans* et al., 1976]. Developing embryonic pancreas might be taken as a model for studies of organogenesis and differentiation using various substances. It is suitable for a study of the selective effect of certain substances in the formation of specific cells associated with the differentiation process. In the pancreas, dexamethasone selectively increases amylase RNA concentration while BrdU decreases the synthesis of all specific RNA in this organ [*Githans* et al., 1976]. If one admits that an early secretion takes place in endocrine cells and that secretion is a selective process, it becomes possible to select cells specifically for certain proteins and to study their appearance in the environmental medium during development [*Rutter* et al., 1978]. Embryonic pancreas was taken from different developmental stages. Radiolabelling techniques, DNA:RNA hybridization and translation of mRNA into proteins permitted to measure the synthesis of a small quantity of amylase mRNA. This enzyme is not present 14 days after gestation, and its amount increases from 0 to 36 units between the 15th and 20th days of gestation. In adults, the amount of amylase mRNA has been increased 300-fold.

The development of embryonic pancreas in a tissue culture is very similar to that observed in vivo. A pancreas explant before cytodifferentiation into specific cells has been maintained in the presence of BrdU. In treated explants, dexamethasone induces the increase of pancreas-specific proteins from 40% in controls to 70% in treated cases. The amount of amylase is 4 times higher, that of amylase mRNA is increased 2.4-fold. In contrast, explants incubated in the presence of BrdU synthesize practically no exocrine protein or insulin, while DNA and total protein synthesis are slightly inhibited. These tissues contain about 3% of amylase in comparison with that found in controls. The same observation was reported for amylase mRNA in treated tissues. Thus, BrdU appears to modify the affinity of certain proteins involved in the regulation of DNA replication. The process of release of specific information is thus upset.

The tubular gland cells in the chick oviduct magnum, stimulated by estrogen, synthesize and secrete large amounts of the major egg white proteins (ovalbumin, conalbumin, lysozyme and ovomucoid). When estrogen is withdrawn, the synthesis of egg white proteins declines rapidly. Secondary stimulation also takes place with progesterone and glucocorticoids and with steroids that are inactive as primary stimulants [*McKnight and Palmiter*, 1979]. Expression of the ovalbumin gene in chicken oviduct explant cultures requires estrogen and a somatomedin-like peptide hormone while that of the conalbumin gene is fully induced by estrogen alone [*Evans* et al., 1981]. The mechanism was not elucidated. Since estrogen induces local DNA strand separation [*Beljanski* et al., 1981a], the conalbumin gene may be thus liberated while for the ovalbumin gene a somatomedin-like peptide hormone is required in addition.

Stimulation of RNA synthesis is not an exclusive privilege of the steroid hormones. The thyroid hormone stimulates de novo the synthesis of growth factor in GH-1 cells cultured in vitro. Increased synthesis takes place through a protein receptor in the nucleus [*Samuels and Shapiro,* 1976] followed by a rise of mRNA, the main limiting factor. It was also shown that *l*-triiodothyronine induces the accumulation of RNA which precedes growth hormone synthesis of the thyroid hormone. These observations illustrate the fact that release of information and control of gene expression are fundamental problems, the multiparous mechanism of which should receive greater consideration. These detailed examples with dispersed cells in culture or tissue explants, both of which are conditions in which interactions are properly controlled, allow one to have some idea of what is going on in the intact organism.

In view of the fact that steroids and other hormones stimulate RNA synthesis, several authors suggested that the activity of the DNA template, causing RNA synthesis, was increased by hormones. This suggestion was shown correct when it was demonstrated that elevated steroid concentrations act by separating in vitro DNA double strands [*Beljanski* et al., 1981a]. Under physiological conditions, this must take place when the hormone is attached to its protein receptor or when it is dissociated from the complex. Interaction of the 'hormone-receptor' complex with DNA must result in physicochemical modifications of DNA since a sudden sharp increase in DNA and RNA synthesis occurs.

Hormones and the Release of Information in Cancer Cells

Since steroid and peptide hormones cause normal target cells to accelerate DNA and RNA synthesis [*Rousseau* et al., 1973; *Munck and Foley,* 1979], it was important to study the effect of glucocorticoids and steroid hormones in the release of information in cancerous cells or tissues. In this connection it should be noted that the action of hormones can be easily studied by the in vitro physicochemical changes of DNA isolated from cancer tissues [*Beljanski* et al., 1981a].

Hormone-dependent mammary tumors, induced in BALB/c mice by a carcinogen [DMBA = dimethylbenz(a)anthracen], contain very little or no casein mRNA [*Rosen and Socher*, 1977]. Formation of this mRNA is in correlation with hormonal regulation of the mammary gland [*Pauley* et al., 1978].

In the same mouse line, which has specific mammary lesions (hyperplasic alveolar nodules, HAN), the production of casein mRNA depends on the presence of lactogen hormones (glucocorticoids and prolactin). In HAN mice the amount of mRNA increases considerably when mice are treated with a mixture of glucocorticoid and prolactin [*Pauley and Socher*, 1980]. The level of this RNA can be measured by molecular hybridization with DNA, synthesized in vitro, using casein mRNA as template. In contrast, in mammary tumors induced by the carcinogen DMBA, the amount of casein mRNA hardly varies, even after administration of the hormone to the tested mice. It is quite conceivable that the differences observed could depend on the degree of DNA strand opening. It seems to us that in these carcinogen-treated cells, glucocorticoid and prolactin, for some unknown reasons, cannot open or maintain separated strands, containing the casein gene, in a condition which would allow casein synthesis on a large scale. In fact, we have shown that under certain experimental conditions, there is an in vitro additive effect of DMBA and steroid hormones on cancer DNA strand separation [*Beljanski* et al., 1981a]: there exists a threshold, beyond which replication will overcome transcription and selective release of information will be slowed down.

None of the results indicates whether a population of cells responsible for casein mRNA synthesis, which is apparently hormone-dependent, increases in the presence of glucocorticoids or is just maintained, or even decreases in tumors induced by DMBA. The response of tumor cells to hormone action probably depends in part on the nature of the tumor cell itself. In fact, glucocorticoids or dexamethasone cause the synthesis of a growth hormone mRNA without stimulating in any significant manner total RNA synthesis in hyperplasic tumor cells of the rat [*Johnston* et al., 1978]. 1 h after tumor cells of the rat hypophysis were treated by dexamethasone, chromatin was isolated and its properties were studied. It was shown that template activity is significantly increased for RNA polymerase and the number of initiation sites for this enzyme, controlled in vitro by the hormone, also increases. In conditions in which growth hormone synthesis is not induced because of a confluent state of cells, dexamethasone reduces, rather astonishingly, the number of initiation sites for DNA-dependent RNA polymerase. This, in fact, should not surprise us: physicochemical modification of DNA by dexamethasone must depend on the physiological state of the cells which determines the binding sites for this hormone. We do not known if the division of cells is required for its action.

Dexamethasone induces differentiation of the myeloid leukemic cells of mice which are not target cells of hormones [*Honna* et al., 1979].

Lastly, it should be noted that glucocorticoids, which are cytotoxic for mice lymphoma cells, strongly decrease the number of initiation sites for RNA polymerase, while in mutant cells (which do not possess protein receptors for glucocorticoids) this apparently does not happen [*Johnston* et al., 1978]. Overactivation of genes by steroid hormones or ATCH induces lymphoid tumors in mice [*Silberberg and Silberberg*, 1955; *Weir and Malher*, 1953].

RNA Induction by Steroid Hormones in Non-Target Tissues

According to classical data, steroid hormones seem to exert their effect on chromosomes without directly interfering in DNA synthesis [*Gorski and Gannon*, 1976; *Epifanova*, 1966], except in tissues which synthesize protein receptors specific for the binding of steroids. In theory, cells which do not possess these receptors cannot respond to steroids and do not permit mRNA synthesis when submitted to the action of these hormones. However, it has been shown that estrogen may induce ovalbumin mRNA synthesis in differentiated tissue which is not a target for estrogens. This was demonstrated by hybridization techniques using a DNA clone, specific for ovalbumin mRNA [*Tsai* et al., 1979]. This mRNA has been detected in the nuclei of the following tissues: liver, spleen, brain and chicken heart. This RNA possesses sequences corresponding to ovalbumin DNA. Gene expression for the formation of this RNA is estrogen-dependent. The amount of ovalbumin mRNA is of the order of 0.2–0.7% of all mRNA per cell. In control cells untreated with the hormone, this level is less than 0.01%. A significant amount of ovalbumin mRNA has also been found in liver and brain polysomes. It is possible to translate this RNA into protein. Thus, some cells from liver, spleen, heart and brain respond to estrogen action and produce ovalbumin mRNA and ovalbumin. This induction of gene expression remains low before hormone administration but is amplified by a factor of 20 after the action of the estrogen. It appears that the number of unexpressed cells which possibly remain undifferentiated is so small that their presence cannot be accurately revealed without amplification by hormone action [*Caplan and Ordahl,* 1978].

These results may be compared with those obtained when steroids induce differentiation of erythroleukemic cells cultured in vitro [*Friend and Freedman,* 1978]. Could it be imagined that beyond a certain threshold of deregulation a non-target hormone cell may exhibit some sort of response to hormone stimulation? Several consecutive injections of estradiol dipropionate to Singi fish *(Heteropneustes fossilis)* led to increased protein and RNA contents of liver and protein content of plasma of male and female fish. These changes were not observed with ovary or testis and muscle. Testosterone injections failed to cause any of these changes. The DNA content of liver, muscle and ovary or testis was not affected by any of these hormones [*Medda* et al., 1980]. These data show that the effect of estradiol dipropionate resulted only in transcription and translation processes.

Administration of prednisolone or another corticoid over a long period of time to BALB/c mice leads to changes in the lymphoid tissue and modifies the response to infection by Moloney leukemia virus (MLV). Thus, in prednisolone-treated mice infected by MLV, lymphoid tumor or granulocytic leukemia cells appear in large numbers [*Abelson and Robstein,* 1970]. When the virus alone is administered, only lymphocytic leukemia is observed. As far as our concept is concerned, we think that here again, during simultaneous administration of virus + prednisolone, the hormone probably contributes to DNA strand separation. The liberated DNA strand allows granulocytic leukemia to appear and the virus may contribute in increasing or at least maintaining the zones of DNA single strands [we verified that several corticoids are capable of in vitro DNA strand opening; *Beljanski,* unpublished results].

Depending upon glucocorticoid concentrations and the cell type used, this hormone may either stimulate the proliferation of normal cells cultured in vitro [*Gaffney and Pigott,* 1978; *Trash and Cunningham,* 1973] or inhibit it [*Castellano* et al., 1978; *Jones* et al., 1978]. Steroid hormones stimulate the differentiation of various tissues by inducing the accumulation of certain proteins and organelles which are characteristic of differentiated cells [*Sugimoto* et al., 1974]. When muscle cells, isolated from newborn rats and cultured in vitro, are submitted to the action of dexamethasone, one can see the stimulation of myoblast proliferation with no increase in cell differentiation [this latter can be evaluated by measuring the activity of creatine phosphokinase, a protein which accumulates in differentiating cell; *Guerriero and Florini,* 1980]. Muscle cells are not hormone target tissues and, therefore, the action of dexamethasone cannot be explained in the same way as in target tissues, even if non-target organs do contain a limited number of hormone receptors, as we shall see.

It is relatively frequent to find tissues which are not hormone target tissues but possess receptors for glucocorticoids [*Jensen,* 1978; *Hadjan* et al., 1974; *Groudin and Weintraub,* 1975; *Humphries* et al., 1976]. The relative frequency of such examples requires a full explanation. In fibroblast cells from the liver and brain cultured in vitro, small amounts of globin mRNA have been detected, although this tissue is not erythropoietic. [*Humphries* et al., 1976; *Ono and Culter,* 1978]. Proteins such as actin and myosin are found in practically all cells from eukaryotes [*Pollard and Weihing,* 1974] as well as in neuroblastoma cells [*Rein* et al., 1980]. Does a hormone, by its interaction with DNA, force a gene to synthesize something which was not in its programme in the absence of the hormone? It appears that hormones may somehow release a silent message from the genome.

Erythropoietic cells from mice and rats contain receptors of testosterone. This hormone is not metabolized by cells [*Hadjan* et al., 1974]. Testosterone added to an in vitro culture of erythroid cells causes a 3-fold increase in the amount of mRNA. The sedimentation coefficient of this RNA is close to that of globin RNA, i.e., 9S to 10S. This fact prompted authors [*Congote and Solomon,* 1975] to suggest that steroid-induced RNA would be in fact globin mRNA. In vitro translation of such mRNA should confirm this suggestion. The mechanism by which testosterone acts in erythroid cells is not clear.

The amount of mRNA specific for basic secretory proteins decreases 10- to 20-fold after castration of the rats, but administration of testosterone rapidly and substantially reestablishes a practically normal level of specific mRNA in the seminal vesicle [*Hennings and Boutwell,* 1970]. Testosterone may act at the level of synthesis or at the level of RNA maturation in the presence of an RNase. It should be noted that prednisone and 9-fluoroprednisolone induce in vivo an increase of RNase activity [*McLeod* et al., 1963; *Albanese* et al., 1972]. RNases lead to the appearance of RNA fragments which, depending on the RNase used, may act either as primers for DNA replication [*Beljanski* et al., 1975] or as inhibitors of tumorous cell development [*Le Goff and Beljanski,* 1979]. In addition, some small-size RNAs, isolated from calf liver, kidney or lung, stimulate in vitro RNA synthesis in chromatin from various tissues with a relative specificity [*Kanehisa* et al., 1974]. The specific mRNA which is the product of this reaction was not analyzed, which makes it difficult to understand these results.

Hormonal RNA in the Release of Information

Uterine mRNA isolated from rats treated with estradiol may be used to replace the hormone and to induce the same response when injected into the uterus of ovariectomized rats. This RNA, called hormonal RNA [*Trachewsky and Segal,* 1968], may act efficiently in various endocrine glands. Although at first disputed, these results were subsequently confirmed [*Fend and Villee,* 1971; *Hubinout* et al., 1971]. This RNA, which contains a base composition different from that of other RNAs, appears to be the result of a specific response induced by estrogen [*Trachewsky and Segal,* 1968]. Its synthesis is under hormonal control. Its stimulatory effect is inhibited by RNase [*Villee and Goswani,* 1973]. Hormonal RNA appears to be a unique component of endocrine glands, and its role would be to replace hormones and increase the synthesis of enzymes which should synthesize hormones. Its activity is inhibited by cycloheximide, an inhibitor of protein biosynthesis. The progesterone-receptor complex reacts with sites on chromatin (nuclei from chicken, pigeon and Xenopus tissues) to lead to avidin production in the nuclei of the oviduct. In view of the fact that hormone is not necessary, hormonal RNA provides conditions for avidin synthesis. This RNA is interchangeable for several species [*Segal* et al., 1973].

The biological activity of a particular RNA under estrogen control is expressed by hyperplasia of uterine epithelial cells. This effect is abolished by RNase [*Galand and Dupont,* 1973].

In addition to their specific effect in the synthesis of particular proteins, estrogens administered to mice over a long period cause mammary carcinoma much later. The cells of the carcinoma synthesize certain proteins in excess, such as α-fetoprotein [*Abelev* et al., 1963]. It appears therefore that steroids or gonadotropins induce the release of information which is different from information relating exclusively to hormonal RNA synthesis.

Competition between Hormones and RNA Primers

It may be possible to visualize a competition between steroid hormones and RNA primers in DNA replication or transcription. In fact, we have shown that in vitro stimulatory action of RNA primers in breast cancer, i.e. DNA replication, may be suppressed by steroid hormones [*Beljanski,* unpublished results].

To prevent the RNA primers from acting in DNA replication it would be tantamount to assisting RNA polymerase to transcribe this region of DNA into RNA. Arrhythmic replication of DNA would create conditions for specific transcription.

Does the affinity of hormones for a given segment on DNA from target tissues disturb the affinity of other segments in DNA? In other words, is there competition for sites on DNA between steroid hormones, some carcinogens and certain RNA primers required for DNA synthesis? This question has been studied in an in vitro system by using, in particular, cancer DNA isolated from human breast cancer tissues [*Beljanski* et al., 1981a]. When used separately, each of the cited agents stimulates the in vitro synthesis of breast cancer DNA and only slightly that of DNA from corresponding normal tissues. Steroid hormones exhibit a very slight effect on the synthesis of DNA originating from steroid-non-target tissues. Thus, by using breast cancer DNA whose synthesis is strongly inhibited by high concentrations of actinomycin D, we provided evidence for competition between steroids, carcinogens and RNA fragments. The addition of estrone or estradiol ($1 \times 10^{-4} M$) totally suppresses the inhibitory effect of actinomycin D on DNA synthesis. Actinomycin D binds to G nucleotides in DNA, and this same nucleotide is required for steroid hormone fixation to DNA. Our preliminary results show that, at the concentrations used, steroid hormones restore replication of DNA, neutralizing the effect of actinomycin D. Purine-rich RNA primers also neutralize the inhibitory effect of actinomycin D. It indicates that binding sites for the compounds cited are very close together or even identical. It is also possible that, by causing DNA strand separation [*Beljanski* et al., 1981a], steroid hormones enable DNA polymerase to liberate actinomycin D.

Summary

It is commonly considered that steroid hormones exhibit their specific action in target cells containing protein receptors which bind hormones. The steroid-receptor complex, formed in the cytoplasm, is transferred to the nucleus of cells where receptors alone destabilize chromatin, thus allowing transcription of DNA-specific segments by RNA polymerases. Although it is generally believed that cells which do not contain hormone receptors cannot synthesize steroid hormone-dependent specific proteins, some authors have postulated the binding of these hormones to single-stranded DNA originating from target tissues. Others demonstrated stimulation of in vitro synthesis of DNA as well as nucleolar and nuclear RNAs in the presence of steroid hormones, We have reported that steroid hormones strongly stimulate both in vitro synthesis of DNA isolated from hormone target cancer tissues and of DNA from some other tissues. The amount of synthesized DNA appears to be proportional to the degree of DNA strand opening caused by these same steroid hormones.

Steroid hormones compete for binding sites on DNA either with exogenous molecules, such as carcinogens, or with actinomycin D. These substances are also capable of separating DNA strands. Competition for binding sites can be detected between steroid hormones and particular agents, some of which can close DNA chains, especially cancer DNA strands. Steroid hormones active with DNA from target tissues of eukaryotes may directly act on DNA strand opening, particularly on DNA isolated from cancer tissues. Similar results were obtained with plant hormones and DNA from plant tumor cells. There is an additive effect on DNA synthesis in vitro in the presence of different substances generally active on DNA strand opening.

Thus a direct correlation may be established between the amount of synthesized DNA and the extent of DNA strand separation. Both phenomena are linked. Hormones, carcinogens or some other compounds 'activate' in vitro as well as in vivo DNA synthesis, creating single strands in the DNA molecule. Substances capable of closing DNA strands inhibit DNA and RNA synthesis. These two processes are included in the 'activation' or 'inactivation' of genes. Exogenous and endogenous molecules govern these phenomena.

4. Exogenous RNAs in Gene Expression and Transformation of Cells

Introduction

If one supposes that the regulation of gene activity in the cells of higher organisms takes place in the same way as in bacteria, transformation of nuclei would be entirely reversible. Such a regulation does not take place at any stage in mammalian species, but perhaps in Xenopus [*Briggs and King,* 1952]. A theory of gene regulation in higher cells has been proposed [*Britten and Davidson,* 1969]. It is strictly limited to the regulation at the level of genomic transcription and implies the interference of 'activator RNA' in the nucleus. According to this theory, the 'productor' genes of mRNA are activated by 'activator RNAs', these latter having been synthesized on activator genes. The role of activator RNA is to induce the transcription of a large number of activator genes in response to a simple molecular stimulus. Numerous chemical molecules, hormones, steroids or peptides, plant hormones, and inducers from embryonic cells have been detected in the nuclear apparatus of target cells [*Britten and Davidson,* 1969; *Liao* et al., 1966; *Wessels and Wilt,* 1965] where they would act on 'sensor genes' which perform certain functions of the productor and integrator genes. However, the model based on gene activation by activator RNAs at the level of regulation throws no light on the irreversible modification linked to the metabolism of cells undergoing differentiation. There is no evidence that different molecules differing in their chemical nature act exclusively on 'sensor genes'. In addition, the machinery for the production of small-size RNA and oligoribonucleotides is available to the cell. Cellular RNases appear to be essential for the production of oligoribonucleotides, upon which may depend a particular cell function in gene activation, transcription or translation. Depending on necessity, these enzymes may splice large or small RNAs, producing RNA fragments which are more or less rich either in purines or pyrimidines, so that the resulting RNA fragments may be further and slightly adjusted for a specific and given function. How is it possible to induce activation of various genes by RNA?

In order to be functional, a double-helical segment of DNA must be 'activated'. This event depends on numerous factors. We have already seen in the previous chapter that various substances may selectively open a part of the double-helical DNA structure in order to 'liberate' a defined DNA segment. Small-size RNAs and oligoribonucleotides are among the substances that might participate in this process. They also play a fundamental role at the level of mRNA translation into proteins. Depending upon their origin, size and base composition, small-size RNAs and oligoribonucleotides can act either as primers for DNA replication or as 'activators' or 'inhibitors' of protein synthesis. Numerous experiments have also shown the capacity of different exogenous large and small RNAs to transform eukaryotic or prokaryotic recipient cells.

Here we shall consider several aspects of the multiple and leading role played by RNAs in the division, differentiation and transformation of cells. All cells of the metazoan organisms (a pluricellular organism) contain the same set of genes [*Davidson*, 1968; *Gurdon*, 1962] although they synthesize different mRNAs, depending upon the type of tissues they are intended to organize. During cell differentiation which appears in the course of development, transcription of DNA into mRNA should proceed without preventing the total replication of DNA upon which cell division is dependent. This double event is accompanied by an enrichment of cells in specific metabolic activities which will be maintained in differentiated cells except when exogenous molecules intervene to change this state. Beside information emerging from DNA, cytoplasmic and environmental informations are necessarily involved in these events. They influence or even determine the differential replication of DNA in the nucleus and might also be involved at the level of DNA transcription into RNA [*Chandebois*, 1980].

Preexisting mRNA and Translation in situ

Masked and specific mRNA, including histone mRNA, have been found in the nuclei of vertebrate animals [*Hough and Davidson*, 1972; *Raff* et al., 1972]. It has been shown that in unfertilized sea urchin eggs most mRNAs escape active translation, although the translation of some histone mRNA is not negligible [*Gross and Gross*, 1973]. This has led to the conclusion that the egg contains at least half of the amount of mRNA specific for histones [*Kedes* et al., 1969]. From the cytoplasm of a nonfertilized egg, it is possible to isolate

9S RNA associated with mRNA, which is in vitro translated into the specific histones of chromatin [*Gross and Gross,* 1973]. According to certain authors, the distribution of mRNA in the egg could be homogeneous. Others claim that this distribution is not uniform in the cytoplasm and this inequal distribution could be strengthened after egg fertilization. During egg cleavage, the blastomers could each contain different mRNAs so that synthesis of different proteins would be possible: this is the basis of differentiation. It appears that after early synthesis, some proteins are capable of regulating the replication and transcription of DNA during embryogenesis [*Gross and Gross,* 1973].

When one blocks the radiolabelling of nuclear and cytoplasmic RNA of sea urchin eggs with actinomycin D (inhibitor of RNA synthesis) or ethidium bromide (inhibitor of DNA synthesis), one observes that the incorporation of amino acids into proteins takes place normally at least during the first cycle of egg cleavage [*Greenhouse* et al., 1971]. This observation indicates, on the one hand, that the cytoplasm contains, or appears to contain, functional mRNA in the presence of actinomycin D and, on the other, that ribosomes are also present in the cytoplasm. Actinomycin D does not prevent the incorporation of amino acids into proteins of the cytoplasmic microtubules in the unfertilized sea urchin egg, which is a good example of the existence of specific RNAs in sea urchin eggs [*Raff* et al., 1971]. In addition, it has been shown that protein biosynthesis of microtubules does not need the presence of nuclei but remains dependent on ribosomes. An electrophorogram shows no difference between the synthesized proteins in one half of a sea urchin egg, treated with ethidium bromide, and the other, untreated half [*Raff* et al., 1972]. Thus, synthesis of new RNAs does not seem to be necessary for protein synthesis. It should be noted that protein biosynthesis rapidly increases after fertilization of eggs in the absence of synthesis of new RNA including the mRNA of microtubular proteins. These examples and many others [*Crippa* et al., 1967; *Davidson and Hough,* 1971] indicate the existence of 'masked mRNA' whose presence in the unfertilized egg can no longer be doubted [*Raff* et al., 1972]. The presence of amino acids is necessary for the translation of the preexisting mRNA into proteins. The occurrence of masked and protected mRNA during the early life of the embryo and in the cells of differentiated organisms [*Gross and Gross,* 1973; *Brown and Littna,* 1966] has been confirmed [*Gross and Cousineau,* 1969; *Spirin,* 1969]. In young hamster kidney cells cultured in vitro, 'unmasking' of ornithine decarboxylase mRNA has been obtained in the presence either of actinomycin D or diethylstilbestrol (synthetic hormone) or steroid hormones [*Lin* et al., 1980]. Although actinomycin D inhibits the synthesis of RNA, it permits the formation of ornithine decarboxylase in these cells. The presence or absence of diethylstilbestrol does not impede the synthesis of this enzyme. Diethylstilbestrol as well as steroid hormones stimulate the activity of the ornithine decarboxylase which, however, does not appear in the presence of cycloheximide. The latter is an inhibitor of protein biosynthesis. The half-life of the enzyme and that of its mRNA is short (10–20 min), but its role in the synthesis of purines is not negligible, especially during development. Thus, substances as different as actinomycin D and natural steroid hormones or chemically synthesized hormones act by a relatively nonspecific mechanism which is, however, important in the translation process. It is conceivable that by separating DNA strands these substances may liberate the message for the synthesis of a particular RNase which, in turn, acts on rRNA in order to produce the oligonucleotides capable of initiating translation.

Cell Differentiation and DNA Replication

In nondividing cells, DNA replication is not a necessary process for the renewal of certain proteins while transcription is essential. One measures the specific transcription through the appearance either of mRNA (DNA-RNA hybridization technique) or of specific proteins. Is the differentiation correlated with total replication and cell division during embryogenesis? When presumptive myoblasts of chick embryos are cultured in vitro, syncytial polynuclear fibres appear in the daughter cells but only if DNA is entirely replicated. This process is followed by mitosis [*Stockdale* et al., 1966; *Bischoff and Holtzer*, 1968]. The differentiated cells are characterized by myosin, actin and tropomyosin synthesis. These authors maintain that the installation of a specific metabolism could not take place without cell division, which would require total duplication of the DNA. However, it was reported that actin and myosin preexist in the presumptive myoblasts and even in cells which do not produce muscles [*Sturgess* et al., 1980; *Perlman* et al., 1977]; synthesis of these two proteins will be simply amplified. We should recall that initiation of DNA replication requires the presence of RNA primers and that translation of mRNA into proteins needs oligoribonucleotides to initiate or inhibit this event.

Immobilized presumptive myoblasts in phase M for 15 min in the presence of vinblastine or colchicine do not synthesize myosin. Differentiation restarts after the cells have been washed. However, no information was forthcoming about what is going on at the level of specific mRNA. Washing of cells certainly removes a large amount of vinblastine or colchicine, both being fixed to DNA, but small residual concentrations of these drugs could play a role in differentiation, as was shown for actinomycin D or other substances [*Lotem and Sachs,* 1979]. In fact, actinomycin D, considered as an inhibitor of RNA synthesis, induces in vitro differentiation of leukemic cells into macrophages only if its concentration is very low (2 ng/ml). At high concentrations, differentiation is not observed. These results clearly illustrate that an agent, depending on its concentration, may play a role either as activator or inhibitor of cell differentiation. It should be noted that low doses of a carcinogen or an antimitotic locally induce DNA strand separation while high concentrations have no effect. Thus it seems possible to explain the activating or inhibiting effect of this kind of substance in cell differentiation and/or division.

Presumptive myoblasts, incubated in vitro in the presence of DMSO (dimethylsulfoxide, a teratogen compound), synthesize DNA and undergo cell division without the appearance of cell differentiation. Under these conditions, neither myosin nor actin synthesis occurs. If these proteins and their corresponding mRNA preexisted in the myoblasts, a certain amplification in the synthesis of these proteins should be expected. This appears not to be the case [*Holtzer* et al., 1973]. In these experiments, complete replication of DNA and cell division are not the obligatory conditions for differentiation. After washing of myoblasts to remove DMSO, differentiation is resumed rapidly. If cells are incubated in the presence of BrdU, DNA synthesis takes place but differentiation does not occur. It should be noted that neither DMSO nor BrdU are substances naturally present in cells undergoing differentiation. Thus, even if the above observations are useful, they cannot be used to interpret what normally happens during in vivo cell differentiation. DMSO used at concentrations which almost continuously allow DNA strand separation [*Beljanski* et al., 1981a] favors cell division, but impairs cell differentiation in cultured presumptive myoblasts [*Holtzer* et al., 1973].

Along these lines it is worthwile to recall that in the thymus of corticosteroid-treated chick embryos DNA polymerases and activities were markedly decreased, 2 days after dexamethasone treatment. The recovery of these enzymes was total at birth [*David* et al., 1980]. These data indicate that DNA synthesis may be impaired by transcription of some specific genes.

In the embryos of amphibians, the doubling of the quantity of synthesized RNA corresponds to the doubling of the amount of DNA [*Brown and Littna,* 1966]. In fact, each type of RNA doubles in the S phase, during which most DNA is replicated. The accumulation of high amounts of mRNA before differentiation indicates that transcription takes place before replication.

Small RNA in Translation;
Relationship with Release of Information

The presence and interference of various proteins during DNA replication and transcription [*Herrick and Albertz,* 1976; *Franze-Fernandez and Pogo,* 1971] suggest a relationship between replication, transcription and translation. This relationship, as we shall try to demonstrate, may depend on oligoribonucleotides. It is obvious that differentiation of dispersed cells cultivated in vitro is never an exact image of what is going on within tissues, aggregates or the tissues of an organism in development. However, information obtained using cells in culture or tissues maintained in appropriate conditions may appreciably facilitate a study of the mechanism involved in the in vivo release of specific information. Observations have shown that oligoribonucleotides are necessary both to initiate the replication of DNA [*Chargaff,* 1977; *Beljanski,* 1975] and to translate the mRNA into proteins [*Furth and Natta,* 1972; *Reichman and Penman,* 1973; *Berns* et al., 1975; *Lee-Huang* et al., 1977]. It was therefore conceivable that by their presence oligoribonucleotides or even small-size RNA could modify the rhythm of DNA replication and consequently lead to a desynchronization of mitoses. Less frequently, occurring mitoses are conditions for differential transcription, and this appears to determine the amount of mRNA.

How does an exogenous or endogenous RNA put a specific gene into an activated state? Which kinds of RNAs are capable of doing it? Can these RNAs favor the process of differentiation of specific characters in embryonic cells or cause new characters to appear in the differentiated cells which can then be maintained as such?

We have shown that RNA fragments rich in purine nucleotides (G and A), obtained by mild enzymatic degradation of rRNA with pancreatic RNase, interact in vitro with DNAs from different origins but do not necessarily promote the synthesis of DNA from various other sources [*Beljanski* et al., 1975, 1978b, 1981a]. In vivo, some RNA fragments of that type stimulate the genesis of leukocytes and platelets and direct differentiation of stem cells in such a way as to correct the deficiency induced by toxic substances [*Beljanski* et al., 1978b]. Also, other RNA fragments, obtained by degradation of rRNA with U_2-RNase [*Le Goff and Beljanski,* 1979], may be used to block the division of plant tumor cells. In vitro, these RNA fragments inhibit the DNA synthesis of plant tumor cells without inhibiting that of DNA from healthy cells.

In association with DNA-dependent DNA polymerase and unwinding enzymes, RNA fragments which are *active as primers* for replication of some DNA will accelerate local separation of DNA strands by their attachment to DNA. Their action in gene activation could be explained by the induction of physicochemical changes of one strand. These changes could lead to the appearance of longer single-stranded segments, thus creating active zones which serve as templates for RNA polymerase. Thus, release of specific information could depend on the type of RNA fragments.

Participation of oligoribonucleotides in the translation of mRNA into proteins has been first postulated [*Furth and Natta,* 1972; *Goldstein and Penman,* 1973] and then confirmed by experimental data [*Berns* et al., 1975; *Heywood* et al., 1974; *Bester* et al., 1975].

The existence of small-size RNA in the nucleus of eukaryotes [*Goldstein,* 1976; *Shih and Bonner,* 1969; *Zieve and Penman,* 1976] and of oligoribonucleotides containing various amounts of purines over pyrimidines has also been demonstrated in the ribosomal preparations. Some of these act as initiation factors for protein biosynthesis while others behave as inhibitors.

It would be essential to know the origin of these oligoribonucleotides and to what extent they contribute in maintaining a relationship between transcription and translation.

Small nuclear RNAs seem to be produced during the transcription of certain segments of DNA. Their migration from the nucleus to the cytoplasm and vice versa has been observed [*Rein,* 1971]. However, no hypothesis concerning their action in the cytoplasm has been put forward [*Goldstein,* 1976]. During splicing or degradation of giant RNAs and formation of the informational mRNA, there are 'non-coding RNA pieces' which may be used as sources for the production of a particular type of fragments, oligoribonucleotides. Another source of oligoribonucleotides in the cytoplasm may be rRNA and unstable mRNA, this latter having already been translated into protein. The appearance of oligoribonucleotides may be coordinated with the translation of mRNA, because a cell is an organized unit, of which cell economy is an important part. In this connection, we should recall that poly(A) residues (20–150) attached to informational mRNA may, after translation, be the target of a nuclease in order to produce oligoadenylates which, in turn, may inhibit protein biosynthesis as was observed with 2′,5′-oligoadenylates [*Kerr and Brown,* 1978]. The inhibition of protein synthesis could probably take place through mRNA degradation by a nuclease. This interpretation is supported by the observation that cells treated with interferon produce an RNAse which degrades viral mRNA and the same RNase could be produced also by 2′,5′-oligoadenylates [*Ball and White*, 1979; *Wreschner* et al., 1981]. Similar results were obtained with human lymphoblastoid cells treated with glucocorticoids [*Krishman and Balgioni*, 1980].

The initiation factors required for protein biosynthesis are complex molecules which lose their activity if they are dissociated into oligoribonucleotides and protein fraction. The presence of oligoribonucleotides is an obligatory requirement for the restoration of the activities of initiation factors involved in protein synthesis. Such an active oligoribonucleotide, called 'iRNA' [*Berns* et al., 1975], rich in adenine (46%) and poor in guanine (7%), has been isolated from a reticulocyte ribosomal preparation, but is also present in ribosomes from other sources. In vitro 'iRNA' is used for translation of different mRNAs. Does this happen in vivo? No data have been reported on the subject as far as we know.

Another type of small RNA has been found in the dialysate of chick embryo muscles and termed 'translation control RNA' (tcRNA). It inhibits in vitro the translation of heterologous mRNA but does not act on that of homologous mRNA [*Bester* et al., 1975].

Two types of oligoribonucleotides have been isolated from *Artemia salina* embryos [*Lee-Huang* et al., 1977], activator RNA, abundant in dormant embryos and inactivated by RNase T_1, and inhibitor RNA which loses its activity after treatment with pancreatic RNase. Thus, hydration of an RNase may lead to the appearance of enzyme activity upon which will depend the amount and nature of activator or inhibitor oligoribonucleotides and, consequently, the translation process itself [*Lee-Huang* et al., 1977; *Bester* et al., 1975]. It should be noted that these oligoribonucleotides were not tested to verify whether or not they were active in the replication of DNA from homologous or heterologous species. RNAs in the ribosomes contain 200–300 nucleotides particularly exposed to RNase action. This may also account for the appearance of activator or inhibitor RNA fragments involved in the translation events. One should take into account that very many substances (hormones, RNA, ions, some alkaloids) as well as radiations, pH, etc. greatly affect the activity of RNases. It has also been shown that in the presence of ATP, 5S rRNA, which is an inhibitor of in vitro protein synthesis, induces degradation of polysomes by an RNase, bound to ribosomes. In vivo, 5S rRNA would behave like double-stranded polyribonucleotides poly(I)–poly(C) which inhibit protein biosynthesis [*Content* et al., 1978]. Now, we know that certain proteins are necessary for transcription of DNA into RNA.

Double-stranded RNA exists in the cytoplasm of normal cells [*Kronenberg and Humphreys,* 1972] and of those infected with viruses [*Hunt and Ehrenfeld,* 1971; *Goldstein and Penman,* 1973]. They seem to form a complex with the initiation factors in order to prevent the translation [*Kaempfer and Kaufman,* 1973; *Kaempfer,* 1974]. Other small RNAs, isolated from newly born rats, alter translation of mRNA in wheat germ [*Berns* et al., 1975]. Such an RNA, containing poly(U) stretches, inhibits the translation of procollagen mRNA and homologous or heterologous mRNAs [*Zeichman and Brutkrentz,* 1978]. In contrast, RNA without poly(U) appears to activate the translation slightly.

In rabbit reticulocyte lysates, 2′,5′-oligoadenylates, containing essentially three residues and synthesized in the extracts of lymphocytes, inhibit protein biosynthesis [*Kerr and Brown,* 1978]. The presence of oligoadenylates in different mammalian cells [*Stark* et al., 1979] suggests their importance at various stages in cell activity. In chicken oviduct pretreated with estrogen, oligoadenylate synthesis takes place as soon as the hormone injection ceases. This latter controls the synthesis of specific mRNA [*Shimizu and Sokawa,* 1979]. The transformation of blasts by phytohemagglutinin [*Howanessian* et al., 1979] may be inhibited by oligoadenylates. These oligoribonucleotides and several RNAs involved in the release of information through translation are largely interdependent since they are indirectly regulated by RNases which are capable of producing different oligoribonucleotides.

Giant Nuclear RNA as Source of RNA Fragments

In eukaryotic cells with genomes 100–1,000 times longer than in prokaryotes, the nucleus is almost completely separated from the cytoplasm by a membrane [*Chandebois,* 1976]. Transcription of DNA into RNA takes place inside the nucleus (except for mitochondrial DNA) while translation into proteins is carried out essentially in the cytoplasm. Giant RNAs, i.e., precursors of mRNA, are spliced and thereafter various coding segments are ligated in order to constitute the 'mosaic' informational mRNA to which is attached, by covalent linkage, a poly(A) segment [*Philipson* et al., 1971; *Darnell* et al., 1971; *Hyatt,* 1967]. Segments of spliced RNA which do not participate in the formation of coding mRNA (introns) may act either on chromosomes to liberate some regions of DNA or are split into oligoribonucleotides, some of which migrate to the cytoplasm for the control of protein biosynthesis. These oligoribonucleotides and small-size RNAs are copied on DNA, and for this reason they have the privilege of reacting with certain sequences on DNA or even with mRNA. They may spread out from the nucleus to the cytoplasm and vice versa and are excellent candidates for interfering in the activities of cells and their differentiation. Due to their small molecular size (10–80 ribonucleotides), they may, as complementary sequences, interfere as primers for DNA replication. Their affinity for DNA would depend on their base sequence. It is thus conceivable that oligoribonucleotides originating from various sources may also, if they are accepted by cells, interfere in the replication of DNA or participate in the differentiation of a cell population, or simply accelerate differentiation. In fact, it was reported that the introduction of pancreatic RNase in the region of the anterior pole of *Smittia* eggs leads to the formation of two abdomens [*Kandler-Singer and Kathoff*, 1976]. It should be recalled that a model has been proposed [*Britten and Davidson*, 1969] conferring on some 'undefined types of RNA' the role of gene activators and regulators, whithout taking into account either the 'overimpression of genes' [*Grassé*, 1973, 1977; *Barrell* et al., 1976a, b] or the irreversible changes of metabolism during differentiation. If DNA replication were accomplished on successive genes, some RNA primers should be eliminated and degraded. This was, in fact, observed with RNA primers termed BLRs [*Beljanski* et al., 1978b, 1981a, b]. However, after degradation, some di- or tri-ribonucleotides, containing A and G nucleotides in particular, may initiate transcription. Non-coding RNA fragments and those originating during splicing of precursors of rRNA may also furnish RNA primers for DNA replication through RNase action. These events indicate the existence of cell economy which will provide the following components:

(a) RNA primers for DNA replication [*Beljanski* et al., 1975];

(b) RNA regulators of transcription where participation of di- and tri-ribonucleotides is not excluded;

(c) RNA activators and inhibitors acting at the level of translation [*Bester* et al., 1975; *Lee-Huang* et al., 1977];

(d) transforming RNA which may result from degradation of rRNAs and some mRNAs by nucleases [*Beljanski and Aaron-da Cunha*, 1976].

RNA and in vitro Differentiation of Some Tissues

Direct transfer of RNA among cells appears to be possible [*Kolodny*, 1971; *Barghava and Shanmugam*, 1971]. Penetration of viral RNA or cellular DNA into recipient cells has been reported [*Pagano*, 1970; *Paffenholz* et al., 1976; *Niu* et al., 1973]. The penetration of RNA or DNA into the nucleus is also quite possible since informosomes, i.e., large-size molecules, resulting from the association of mRNA and proteins, are known to leave the nucleus and enter the cytoplasm [*Spirin,* 1969].

Exogenous RNA, Inducer of Neural Cells

Cultured in vitro, 'chord' cells release ribonucleoprotein (RNP) particles into the medium [*Niu,* 1956]. Acting on ectoderm, these particles cause the formation of neural and muscle cells in the chick embryo. On the other hand, RNA isolated from guinea pig bone marrow [*Sasaki and Niu,* 1973] and incubated with ectoderm leads to the appearance of neural cells. Brain mRNA initiates chick blastoderm to develop into nervous tissue. RNase and actinomycin D inhibit these cellular manifestations. Do they inhibit the competence of the ectoderm or do they interact with exogenously added RNA? Serum albumin, mixed with RNA, significantly amplifies the induction of neural tissue. However, most preparations of serum albumin, even purified, exhibit RNase activity (also DNase) which is capable of producing RNA fragments from exogenous RNA [*Beljanski,* unpublished results]. It would appear that, by accelerating DNA replication in cultured cells, these latter play a role in the transcription of DNA and, consequently, in cell differentiation. However, some polypeptides participate in the differentiation of dissociated cerebral hemisphere nerve cells in culture [*Sensenbrenner* et al., 1975]. Similar results were observed in other species by increasing the pH value of the incubation medium or by eliminating Ca [*Holtfreter,* 1944]. Competent axolotl ectoderm produces neural tissues even in saline solution [*Barth,* 1941]. In a serum-depleted medium in which the pH value had been elevated (8.2–10) for a few minutes, sparse 3T3 cell proliferation was stimulated. More than 80% of the quiescent cells initiated DNA synthesis 9–12 h after the treatment. Increased DNA synthesis and mitosis were observed during one cell cycle. Several interpretations were proposed, except the effect of alkaline pH on the physicochemical structure of DNA [*Zetterberg and Engström,* 1981]. Hydrogen bonds are disrupted more or less rapidly at alkaline pH. We interpret these results as the consequence of DNA strand separation, causing accelerated replication and transcription and thus leading to increased cell division. An increase in the concentration of salts which bind to phosphates of DNA leads to the stabilization of the DNA double-helical structure, modifying the replication/transcription ratio [*Engel and von Hippel,* 1978; *Beljanski* et al., 1981a]. By penetrating into cells, exogenous RNAs or RNA fragments may form a DNA-DNA-RNA triplex which would result in separation of DNA strands or prevention of DNA or RNA synthesis.

RNA Inducer of Heart Tissue in Chick Embryo

Heart muscle differentiation is an early event in chick embryo development [*Romanof*, 1960] and can be taken as a model for studying the evolution of embryonic cells into highly differentiated cells. Chick blastoderm (0.6 mm behind Hansen's node) has been exposed to the effect of RNA molecules containing heart mRNA [*Desphande* et al., 1973]. These RNAs from the nucleus and cytoplasm of chick heart, as well as RNA preparations containing mRNA [+ poly(A)], induce pulsations characteristic of heart tissue. The pulsations do not appear when the blastoderm is incubated without exogenous RNA or when it is incubated in the presence of RNA isolated from brain tissue. The amount of tissue exhibiting pulsations after incubation with nuclear, cytoplasmic and mRNA isolated from chicks represents 48, 58 and 88%, respectively, in differentiated explants and appears to depend on the mRNA concentration. RNase removes the effect of mRNA.

More recently, it was reported that 7S RNA, when isolated from 16-day-old chick embryo heart [*Siddiqui* et al., 1979], induces changes in the embryonic cells (stage 1) of chick blastoderm cultured in vitro. Pulsating tissues appear, under the conditions described, in a proportion comparable to that occurring during normal differentiation of whole embryos. In addition, the action of RNA is accompanied by the appearance of acetylcholinesterase activity [*Desphande* et al., 1978]. This is abolished by RNase. Here also, RNAs originating from chicken kidney or thymus are inactive, as are synthetic polynucleotides. Although rich in poly(A), 7S RNA does not behave in vitro as an mRNA in an appropriate system [*Siddiqui* et al., 1979]. Its role in heart tissue differentiation is not explained.

However, it may be imagined that the chick blastoderm used contained cells possessing informative molecules and that exogenous RNAs, or their degradation products, have simply accelerated the process of differentiation in some way. If exogenous active RNAs were functioning as such in vivo, the systhesis of specific proteins would take place in order to complete those wich are already existing. The myogenesis implicates important changes in protein biosynthesis, and the transformation of myoblasts to myofibrils [*Paterson and Bishop*, 1977; *Delvin and Emerson*, 1979] necessitates the accumulation of new classes of mRNA. Is there an amplification of an already existent metabolism so that 7S RNA only enhances this metabolism? We do not have the answer. Although the existence of 7S RNA rich in purines has been demonstrated in normal chicken cells used for preparation of viral RNA [*Erickson* et al., 1973], its role has not been elucidated. However, an mRNA appears to be able of inducing the functional intercellular communication in already organized tissue [*Dahl* et al., 1981]. These communication channels may permit different exogenous substances to activate or inactivate particular genes.

RNA as Epigenetic Signal during Development

Induction of embryogenesis requires interaction between cells. One of the fundamental characteristics of cells is their plasma membrane which, through its lipoprotein constituents, may modify the activity of certain genes. Thus, glucose-6-phosphate phosphatase (EC 3139) [*Tice and Barrnett,* 1962] is a multifunctional enzyme, localized in the endoplasmic reticulum [*Lueck and Nordlie,* 1970]. The activity of this enzyme varies during differentiation and development. It is admitted that channels exist between cells which permit the migration of molecules, even extracellular RNA [*Pagano,* 1970], which penetrate into the nucleus of the recipient cells [*Kay,* 1961; *Ledoux and Charles,* 1967; *Watters and Gullino,* 1971; *Barghava and Shanmugam,* 1971], thus modifying the chromosomes [*Petricciani and Patterson,* 1974]. This type of molecules may maintain the cells in a stable state or favor the synthesis of new proteins [*Lowenstein,* 1978]. Direct cell-to-cell contact between epithelium and embryonic mesenchyme permits the circulation of information and results in the capacity of mesenchyme cells to induce differentiation in the adjacent epithelium. Certain authors have postulated that RNA may carry information representing the determinant morphogen in the induction of embryonic cells from heterotypic cells [*Slavkin and Croissant,* 1973; *Niu,* 1956; *Brachet,* 1942]. The role of a particular RNA in the diffusion of information from mesenchyme to epithelium has been postulated [*Slavkin and Zeichmer-David,* 1981]. Indirect evidence suggested that a low molecular weight 'regulatory RNA' may mediate the transduction of information from mesenchyme to epithelia. However, the membrane which separates these two sheets might be an obstacle to the transfer of such an RNA.

Relatively small methylated RNA molecules have been found in the extracellular matrixes interposed between epithelium and the adjacent mesenchyme [*Slavkin and Croissant,* 1973; *Slavkin* et al., 1969]. Incorporation of tritiated uridine in the vesicular matrix [*Slavkin* et al., 1971] has been studied by autoradiography using electron microscopy. Within 15 min, tritiated uridine was localized in the vesicular matrix and in the mesenchymal cells, adjacent to the epithelial cells forming the enamel of the tooth. RNA may circulate from the inducer to the receptor tissue. It appears necessary that recipient epithelial cells contain DNA in the process of replication so as to respond to the inductive signal emerging from the mesenchyme [*Saxén* et al., 1976]. During contact between mice tooth mesenchyme and 5-day-old chick embryonic pharyngeal arch epithelium, the mesenchyme cells induce the tooth form and cause the epithelium to differentiate into secretory ameloblasts [*Kollar and Baird,* 1970]. We think that replication of DNA facilitates interference of an RNA which may bind more efficiently to single- than double-stranded DNA. If RNA prevents the replication of one DNA strand on a given site, transcription may proceed on the other strand, thus leading to the release of specific information.

Transformation of Cells by RNA

The term 'transformation' is defined as a process which, under the effect of exogenous RNA, may oblige differentiated or undifferentiated cells to express new characters transmissible from generation to generation (or maintained in differentiated cells?). Studies in which exogenous RNAs were used to induce transformation of recipient cells are compatible with our knowledge of the influence of cytoplasmic molecules (comprising RNA) on the nucleus during the development of pluricellular organisms [*Morgan,* 1934; *Stedman and Stedman,* 1950]. Freshly fertilized goldfish eggs were injected with various RNAs or DNAs: rRNA and partially purified mRNA from matured carp ovarian eggs, liver, and rat Walker ascites cells. DNA was isolated from carp testis and liver. Carps possess a single caudal fin and goldfish a double fin. This feature develops early, and positive identification can be made in 2–3 weeks. Transformation of the caudal fin was observed in fishes emerging from fertilized eggs injected either with mRNA isolated from matured eggs or DNA from carp's testis and liver [*Tung and Niu,* 1973]. The frequence of the appearance of the single caudal fin in the offspring with mRNA or DNA was 26 and 33%, respectively, while in control fish, which received neither RNA nor DNA, the frequency was of the order of 2.2%. Pretreatment of RNA and DNA with RNase or DNase abolished the transmission of the character. The data clearly reveal that the induced character of a single-tail fin in goldfish has been transmitted to the offspring [*Tung and*

Niu, 1975]. The same results have been obtained with DNA which contained about 2% of RNA. It should be emphasized that liver mRNA, ovary rRNA or tumorous mRNA did not exhibit activity as far as the induction and appearance of a single caudal fin in the progeny is concerned. This does not exclude the possibility of some metabolic changes in certain tissues of the progeny due to the effect of these RNAs.

The induction of a new character in goldfish with active mRNA originating from the matured eggs was interpreted as follows: mRNA would be transcribed into DNA by a reverse transcriptase enzyme known to copy mRNA into DNA and which has been found in chick embryos [*Kang and Temin,* 1972]. The fact that different RNAs are inactive in this respect, except mRNA from matured eggs, indicates that nucleotide sequences of mRNA are of crucial importance and favor the proposed interpretation, since the same results were obtained with carp testis DNA. According to these data, one should find nucleotide sequences complementary to active mRNA in the genome of the offspring of single-tail fishes. However, one may ask if carp mRNA or part of it which is active in the induction of tail transformation in goldfish was first integrated into the mRNA of recipient cells of caudal fin in order to form 'mosaic mRNA' and was then transcribed into DNA, or if it was totally and directly transcribed into DNA without being part of the 'mosaic mRNA' of the goldfish. 'Mosaic mRNAs' are normally produced in lower vertebrates [*Darnell,* 1978]. It remains to be seen how exogenous mRNA controls the expression of a silent gene. Is the integration of exogenous mRNA facilitated by the fact that cells synthesize less RNA at the stage of the caudal fin compared with cells from ectoderm-mesoderm? The latter cells are rapidly dividing, synthesize a great variety of mRNA and give rise to a greater variety of tissues than endoderm cells [*Flickinger,* 1973]. According to this author, formation of various mRNAs would be dependent on DNA replication. In any case, the observation that goldfish eggs injected with mRNA from matured carp eggs give rise to single tails in the offspring, and that this new character is transmitted to the second generation suggests that the information brought by carp mRNA is integrated into the genome. However, the mechanism of this integration remains to be analyzed. It was reported that among different fish egg mRNAs two fractions of mRNA possess the capacity to initiate the formation of liver and red blood cells. Two lines of evidence were presented for this. First, egg mRNAs are capable of encoding the synthesis of albumin and hemoglobin. Second, injection of carp or rat liver mRNA into fertilized goldfish egg resulted in the development of LDH (lactate dehydrogenase), and injection of rabbit globin mRNA leads to the formation of rabbit hemoglobin and rabbit LDH in goldfish red blood cells. These examples show the role of mRNAs in cell differentiation and organ formation in normal development [*Niu*, 1981]. In the process of tail regeneration in the newt *Triturus (Diemictylus) viridescens*, it has been shown that liver RNA has inhibitory effects on the process of regeneration, while RNA prepared from tail tissue increased regeneration over the controls [*Wolsky*, 1974; *de Issekutz-Wolsky* et al., 1966].

RNA Fragments as Promoters of Leukocyte Genesis in Mammals Depleted by Anticancer Drugs

Our biochemical studies on DNA replication led us to prepare several types of short RNA fragments which, depending upon their base composition, possess a large specificity in priming the replication of DNAs [*Beljanski* et al., 1975; *Beljanski*, 1975]. Under suitable conditions, RNA fragments were obtained by fragmentation of purified *E. coli* rRNAs with pancreatic RNase. These RNA fragments markedly stimulate the in vitro replication of DNA isolated from monkey and rabbit bone marrow and spleen. They have little or no effect on the replication of DNA from kidney, muscles, brain, etc. After an intravenous injection of radiolabelled RNA fragments into rabbits, it can be demonstrated that they concentrate in the bone marrow and spleen tissues [*Beljanski* et al., 1978b, 1981a, b]. When healthy rabbits are treated with high dosages of a drug such as cyclophosphamide, a significant decrease in the level of circulating leukocytes is observed. About 24–48 h after a few RNA fragments have been injected to these animals, a normal level of leukocytes is restored and remains within physiological limits (fig. 12). If high concentrations of cyclophosphamide are injected daily, animals can be saved provided that RNA fragments are administered every 5–7 days. Such animals will survive in excellent condition [*Beljanski* et al., 1981a, b]. Similar results have been obtained for the restoration of platelets, the level of which was decreased by drugs (fig. 4). Restoration of leukocytes and platelets is not due to the liberation of these cells from a reserve pool, but to the genesis of new cells through differentiation of stem cells in the bone marrow. Repeated decreases of leukocyte and platelet levels by chemical drugs and subsequent increases of these cells after administration of RNA fragments (called BLR: Beljanski Leukocyte Restorer) can be carried out for a period of several months without exhausting the reserve pool. An excess of BLR does not lead to an excess of leukocytes or platelets because these RNA fragments act in

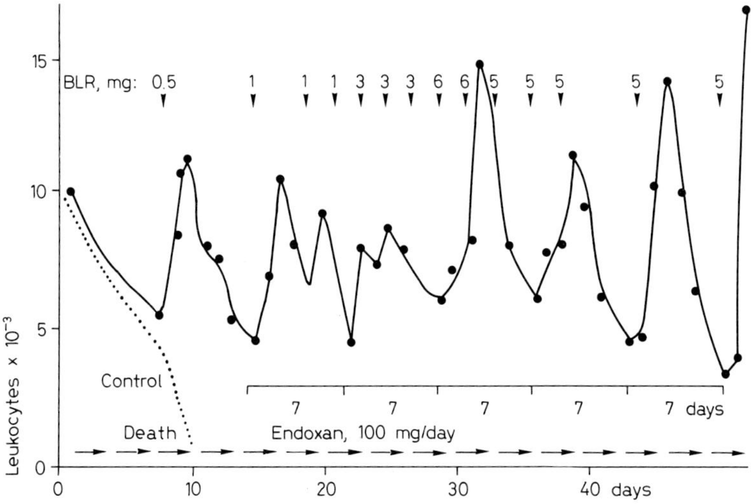

Fig. 12. Leukocyte count in the endoxan-treated rabbits receiving varying intravenous doses of BLR (Beljanski Leukocyte Restorer) every 2nd day. After the leukocyte count had been strongly decreased, a 3.5-kg rabbit treated with Endoxan (100 mg/day) received varying doses of BLR ranging from 1 to 6 mg every 2nd day as shown by the arrows. Circulating leukocytes were counted daily with a Coulter Counter. The results given are an average obtained with 10 rabbits. The mean increase in leukocyte count was 172 ± 17% (standard error). The confidence interval was calculated using paired-sample Student's t test: $p < 0.001$ [from *Beljanski* et al., 1981a].

a physiological manner, as naturally occurring substances in mammalian organisms. The imbalance between lymphocytes and polynuclears may also be corrected, in the first instance, within a few hours through the restoration of the lower type of cells. This indicates the role played by BLR not only in the differentiation of bone marrow stem cells, but also in cells in the multiplying or maturation compartment. This is further illustrated by the strong affinity of RNA fragments for DNA from bone marrow or spleen, i.e., rapidly dividing leukopoietic cells. This may explain the protection enjoyed by leukopoietic cells against drugs which bind to the same DNAs. In humans after severe chemotherapy, BLR-RNA fragments restore normal leukocyte levels (table II). A myelogram taken after chemotherapy of a patient suffering from leukemia shows the presence of many undifferentiated cells with no intermediary cell derivatives in the multiplying compartment. 3 weeks after administration of BLR the same bone marrow shows that undifferentiated cells have disappeared and the bone marrow contains a normal number of intermediary cells in the multiplying and maturation compartment.

Table II. Differentiation of undifferentiated leukemic stem cells in a patient treated with BLR (Beljanski Leukocyte Restorer)[1]

		Before BLR		After BLR
Undifferentiated stem cells		58		0
Myeloblasts		0		1
Promyelocytes		0		0
Myelocytes				
Neutrophils		0		25
Eosinophils		0		0
Basophils		0		0
Metamyelocytes				
Neutrophils		0		11
Eosinophils		0		0
Basophils		0		0
Granulocytes				
Neutrophils	4		12	
Eosinophils	2	6	0	12
Basophils	0		0	
Plasmocytes		2		2
Megakaryocytes		2		few
Proerythroblasts		0		11
Erythroblasts				
Polychromatophils	5		10	
Acidophils	8	13	13	33
Basophils	0		10	

[1] Courtesy of Dr. *J. Bugiel,* Paris [personal commun.].

It is remarkable to observe that in patients with leukemia, BLR protects and restores only normal leukocytes, not the leukemic ones. This is a considerable advantage for the use of BLR since the molecules recognize DNA from normal cells, but not that from cancer cells [*Beljanski* et al., 1981b]. This is one of the evidences of the difference between the two DNAs from normal and cancer cells, as already proven with carcinogens, hormones and alkaloids [*Beljanski* 1979; *Beljanski* et al., 1981a; *Beljanski and Beljanski,* 1982; *Le Goff and Beljanski,* 1981, 1982]. It seems also to indicate that other types of RNA primers, different in size and nature, would be necessary for DNA from cancerous cells, implying modified RNases in cancer cells in order to furnish modified RNA primers. In fact, cancer tissues contain modified RNases, as has already been established in several laboratories [*Reddi*, 1966; *Akagi* et al., 1978; *Kottel* et al., 1978; *Le Goff and Beljanski*, 1981].

Transformation of Oncogenic Agrobacterium tumefaciens *into Non-Oncogenic Bacteria by an RNA from* Escherichia coli

RNAs are capable of inducing transformation of cells from different origins into cells which acquire new characters transmissible to the progeny [*Tung and Niu*, 1975; *Evans*, 1964; *Mishra* et al., 1975; *Beljanski* et al., 1972a, c].

We have shown that an RNA released into the culture medium by showdomycin-resistant bacteria of *E. coli* [*Beljanski* et al., 1971a, b, 1974b], possesses transforming potential when introduced into the medium of certain heterologous (*Agrobacterium tumefaciens*) or homologous (*E. coli*) bacteria [*Beljanski* et al., 1972a, c]. This RNA [6S RNA, rich in purine nucleotides $(G+A)/(C+U)=2.0$] may in vitro be replicated by polynucleotide phosphorylase [*Plawecki and Beljanski*, 1971] and transcribed into DNA by bacterial reverse transcriptase [*Beljanski and Beljanski*, 1974]. Partial transformants of *A. tumefaciens* have lost a part of their oncogenic power, while complete transformants have totally lost this property [*Beljanski* et al., 1972a, c]. In the complete transformants we have found, by hybridization, a copy of 6S RNA as DNA integrated into the genome of transformed bacteria [*Beljanski and Plawecki*, 1973].

The transformed bacteria exhibit the serological properties of *Agrobacterium tumefaciens* [*Beljanski* et al., 1972a], grow faster than wild-type bacteria and synthesize a considerable amount of *l*-asparaginase compared with nontransformed cells [*Beljanski* et al., 1972c]. While the oncogenic strain of *Agrobacterium tumefaciens* contains a very low amount of 23S rRNA which can be isolated, partially transformed strains contain a higher amount, and completely transformed ones possess a 23S RNA in an amount comparable to that found in other bacterial species (fig. 13). This indicates that various transformed cultures contain RNase with modified activity. Neither rRNA nor synthetic polyribonucleotides or a mixture of ribonucleotides are capable of transforming *Agrobacterium tumefaciens*. Transformation was obtained only with 6S RNA, rich in purine nucleotides, which is a type of RNA found in the culture medium of showdomycin-resistant *E. coli* [*Beljanski* et al., 1970] or bound to DNA of these bacteria and is present also in the wild type [*Beljanski* et al., 1974].

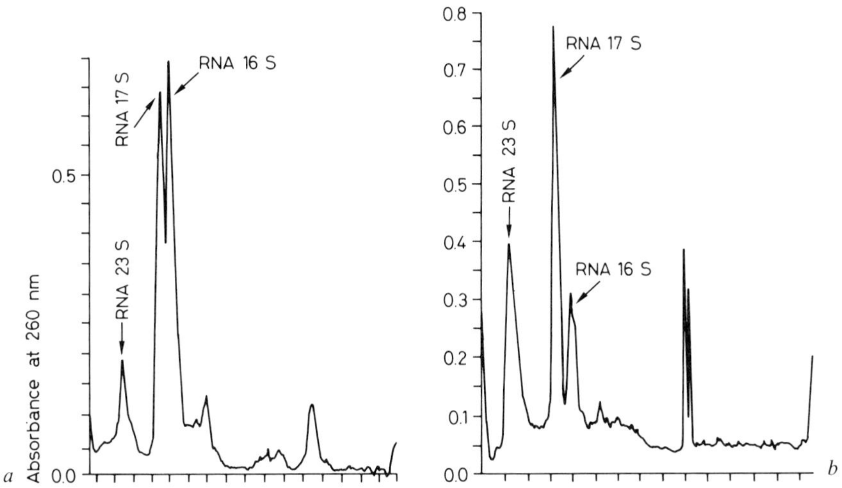

Fig. 13. Densitometer tracings at 260 nm of rRNA from *Agrobacterium tumefaciens* after gel electrophoresis. *a* rRNAs from *A. tumefaciens* B_6 oncogenic strain. *b* rRNAs from partial transformants B_6-Tr-4 obtained by treatment of *A. tumefaciens* B_6 cells by transforming RNA [from *Beljanski* et al., 1972a].

In the course of the transformation of a recipient strain of *E. coli* K-12 by ^{14}C-uracil-labelled transforming RNA obtained from a showdomycin-resistant strain, we have shown that half of the labelled RNA was incorporated by the recipient bacteria. The remaining RNA which did not penetrate the cells was totally precipitable by trichloroacetic acid. After disruption of labelled and washed recipient cells, the membrane fraction and the supernatant contained most of the labelling agent; 5–10% of the incorporated radioactivity was associated with DNA purified by the conventional method. From these data we concluded that radioactive RNA did penetrate the recipient cells [*Beljanski and Manigault,* 1972].

It was reported that wild-type bacteria treated with RNA, isolated from antibiotic-resistant bacteria, have acquired resistance to the corresponding antibiotic. This phenomenon was described for *Pneumococcus* [*Evans,* 1964], *Bacillus subtilis* [*Shen* et al., 1962], and *Escherichia coli* [*Beljanski* et al., 1971a].

An RNA isolated from a wild strain of *Neurospora crassa* which is inositol^{+} is capable of inducing genetic reversion of an inositol^{-} mutant into wild type [*Mishra* et al., 1975]. The specificity of the mRNA from the wild strain, rich in poly(A), was demonstrated by the fact that mRNA isolated from the inositol$^-$ mutant cannot induce the reversion of the locus. RNase abolishes the transforming capacity of the active RNA, while DNase is without effect. Transformed strains, which become inositol$^+$ and appear in a significant number, are stable and transmit the new character to the progeny according to Mendel's laws. It was suggested that information transferred by mRNA to the genome of the transformed strain may be converted to DNA with a reverse transcriptase enzyme. Recently, RNA-dependent DNA polymerase has been found in *Neurospora crassa* [*Dutta* et al., 1977, 1980]. The activity of this enzyme in *E. coli* differs from that exhibited by DNA-dependent DNA polymerases I, II and III [*Beljanski and Beljanski,* 1974].

Transfer of Information from RNA to DNA by RNA-Dependent DNA Polymerase

Besides the DNA-dependent RNA polymerases, the discovery of 'reverse transcriptase' in oncornaviruses has greatly enlarged our knowledge concerning the transfer of information by eukaryotic and prokaryotic cells. In the presence of deoxyribonucleoside-5′-triphosphates, reverse transcriptase catalyzes the transcription of certain RNAs (viral RNA, mRNA and some others) into a complementary DNA [*Temin and Mizutani,* 1970; *Temin,* 1971]. The presence of the activity of reverse transcriptase has been shown also in chick embryos [*Kang and Temin,* 1972], in the microsomes of Xenopus [*Brown and Tocchini-Valentini,* 1972; *Ficq and Brachet,* 1971], in normal human lymphocytes [*Wu and Gallo,* 1975], normal human spermatozoa [*Witkin* et al., 1975], bacteria [*Beljanski,* 1972], monkey placenta [*Mayer* et al., 1974] as well as in plants [*Beljanski* et al., 1974] and fungi [*Dutta* et al., 1977, 1980].

We have presented data on cell transformation under the influence of RNA, and this may be interpreted as resulting from the action of reverse transcriptase [*Beljanski and Beljanski,* 1974]. In certain cases, the existence of complementary sequences of transforming RNA has been detected in the DNA of transformed *Agrobacterium* cells [*Beljanski and Plawecki,* 1973].

By now, numerous examples show that information does not exclusively propagate from DNA to RNA to proteins. The statement that 'it is not conceivable and not proven that information could flow from RNA to DNA' [*Monod,* 1970] is refuted unambiguously by abundant data [*Temin,* 1971; *Wu and Gallo,* 1975].

Induction of Plant Tumors by Small RNA

RNA purified from the tobacco mosaic virus transmits viral disease to healthy plant cells [*Gierer and Schramm,* 1956; *Fraenkel-Conrat* et al., 1961], leading to the appearance of viruses. This implies the synthesis of specific proteins associated with that of viral RNA in healthy plants.

Tumor induction in plants by *Agrobacterium tumefaciens* has raised the question of the tumor principle carried by this species. It was suggested [*Braun and Wood,* 1966; *Stroun* et al., 1971] that an RNA fraction could be an essential factor in plant tumorigenesis. We have shown that a particular (6S) RNA, isolated from bacteria [*Beljanski and Aaron-da Cunha,* 1976; *Le Goff* et al., 1976; *Aaron-da Cunha* et al., 1975], or even some RNA fragments obtained by degradation of rRNA from bacteria, rabbit or monkey liver with pancreatic RNase, induce the appearance of tumors when inoculated into inverted stem segments of *Datura stramonium* maintained under axenic conditions [*Beljanski and Aaron-da Cunha,* 1976]. *Tumor induction requires the presence of the plant hormone auxin and a particular small-size RNA or RNA fragments rich in purine nucleotides.* Purine-rich molecules have not received the attention they deserve, and we will have further occasion to return to this matter. The role of auxin in plant tumorigenesis may well be the opening of DNA strands in *A. tumefaciens* and to a certain extent also in healthy recipient plant cells [*Le Goff and Beljanski,* 1981]. Thus, RNA fragments may either initiate the arhythmic replication of DNA or induce the expression of a silent gene, or inactivate a gene (or genes) leading to dedifferentiation. It has recently been confirmed that an RNA synthesized by *Agrobacterium tumefaciens* is an essential element for plant tumor induction [*Sobota,* 1978]. It should be emphasized that extremely vigorous RNase activity appears in tumor cells, the properties of which are different from those of RNase in normal cells [*Reddi,* 1966]. In addition, it was shown that RNAs isolated from plant tumor cells possess neoplastic activity [*Roussaux,* 1975].

Necrosis of Plant Tumors Produced by Particular RNA Fragments

The common characteristic of oncogenic RNA fragments for plants, obtained by degradation of rRNA with pancreatic RNase, is their richness in purine nucleotides and their ability to act in vitro as primers in the replication of DNA by DNA-dependent DNA polymerase [*Beljanski,* 1975]. We should emphasize that such RNA fragments are not transcribed into DNA

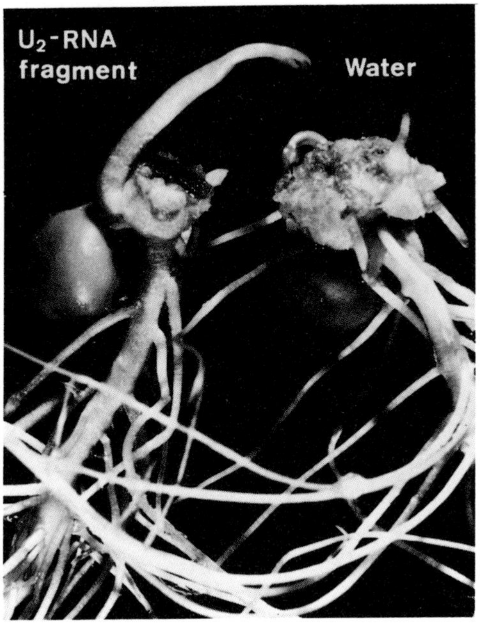

Fig. 14. Inhibition of crown-gall tumor development (day 8) with U_2-RNA fragments (left) [from *Le Goff and Beljanski,* 1979].

under appropriate conditions [*Beljanski and Beljanski,* 1974]. If, however, the same rRNAs are degraded with U_2-RNase (an enzyme which cleaves RNA at A nucleotides), one obtains RNA fragments which are less rich in purines than the rRNA used. In vitro, these RNA fragments act as primers in the replication of certain DNAs, among them plant DNA. In addition, they can be transcribed into complementary DNA by RNA-dependent DNA polymerase (reverse transcriptase), from bacteria [*Beljanski* et al., 1978a, b] or from plants [*Beljanski* et al., 1974]. These U_2-RNA fragments contain 50–65 nucleotides. Depending upon the time they are administered to the wounded plants infected with *Agrobacterium tumefaciens* B_6, they exhibit either a stimulatory action in tumor induction (crown-gall) or act as inhibitors of tumor cell development (fig. 14) [*Le Goff and Beljanski,* 1979]. If placed on the wound of pea seedlings at the same time as the bacteria, U_2-RNA fragments significantly increase the weight of tumors, while administered 24 h after infection with the bacteria (the interval necessary for induction of tumor cells), the fragments are without effect. Applied 4 or 8 days after the bacterial infection, they cause not only a substantial inhibition of tumor cell proliferation, but also large-scale necrosis. In the absence of oncogenic bacteria, U_2-RNA fragments have no oncogenic and no toxic effect on healthy plants. The major inhibition of tumor proliferation in the presence of U_2-RNA fragments is not observed if auxin has been administered at the same time as RNA fragments. This antagonism between auxin and U_2-RNA fragments may be explained by fixation of the U_2-RNA fragments to tumor cell DNA, preventing replication or transcription and resulting in the arrest of cell division. In contrast, auxin actively separates the strands of DNA from plant cancer cells [*Le Goff and Beljanski,* 1981], permitting replication or transcription. Depending on the concentration of auxin, one or the other action will be dominant or suppressed by U_2-RNA fragments.

Our preliminary results have shown that certain U_2-RNA fragments stimulate the in vitro replication of DNA whereas others stimulate transcription [*Beljanski,* unpublished results].

These examples amply illustrate the role of biological molecules such as RNA or hormones in the release or suppression of information from DNA.

Summary

We have collated numerous experimental data to illustrate that depending on the recipient cells, large-size RNA (mRNA), small-size RNA or oligoribonucleotides are capable of inducing cell differentiation or transformation. Thus, specific mRNA isolated from mature carp eggs introduced into fertilized goldfish eggs provokes the appearance of a single tail (a carp phenotype) in the offspring. Both mRNA and 7S RNA, isolated from chick heart tissue, induce heart pulsation in presumptive chick heart explants, which does not appear if the specific mRNA is pretreated with pancreatic RNase or replaced by RNA originating from other tissues. In *Neurospora crassa,* mRNA induces change in a specific gene locus. The essential question which arises from these experiments is to know how mRNA induces genomic changes. One possibility was proposed, i.e., that eggs, embryonic cells, *Neurospora* cells contain reverse transcriptase, an enzyme which transcribes RNA into DNA. This latter could in turn be integrated into the genome of the recipient cell. However, the presence of such a DNA transcript has not been demonstrated. In the case of 6S transforming RNA which produced bacterial variants with stable biochemical and biological characteristics, DNA transcript was detected in the DNA of the transformed strains. The fact that particular oligoribonucleotides are capable of inducing differentiation and multiplication of stem cells from which leukocytes and platelets originate in humans and animals indicates that they interfere with DNA. In fact, they act in vitro as excellent primers for the replication of DNA, isolated from bone marrow and spleen, but are without noticeable effect with DNAs from other tissues. A correlation appears to exist between DNA replication, transcription and translation (at least in dividing cells). We should recall that purine-rich oligoribonucleotides act as activators of protein biosynthesis while those rich in pyrimidine bases act as inhibitors in the translation of mRNA. In the same connection, it should be kept in mind that unfertilized eggs, embryonic cells and differentiated cells contain small amounts of 'protected mRNA'.

5. Carcinogens in DNA Replication and Release of Specific Information

Introduction

For over 10 years, a great many experiments have been providing evidence that carcinogens, antimitotics (most of them carcinogenic), DMSO, phorbol esters and even some solvents may induce cell differentiation. They seem to induce differentiation of cancer cells into normal cells [*Lotem and Sachs,* 1979; *Friend* et al., 1971; *Tanaka* et al., 1975; *Huberman and Callaham,* 1979] and/or a very amplified release of specific information, appearing as mRNA or protein. These studies show that low concentrations of the above compounds may lead to an increased synthesis either of DNA or RNA or even the release of specific information, while higher concentrations may provoke cell degeneration. Carcinogens transform differentiated cells into dedifferentiated ones, i.e., cells which no longer possess the characteristics of normal cell regulation. No precise molecular mechanism has been proposed for these transformations, although the undefined term of 'gene activation or inactivation' is commonly used.

Numerous observations indicate physicochemical changes of chromatin or DNA in normal cells when treated with the cited substances. Present considerations suggest that the ultimate target of carcinogens and carcinogen-like substances is DNA [*Ryser,* 1971; *Shoyab,* 1978; *Beljanski* et al., 1981a]. The DNA sequences, which are carriers of genetic information, do not seem to differ according to normal or cancerous states [*Le Pecq,* 1978; *Sandberg and Sakurai,* 1974; *Myozis* et al., 1980], and this does not support the idea that carcinogens and alkylating agents can distinguish between normal and cancer DNA. In view of this, how can it be explained that carcinogens often induce differentiation of cancer cells maintained in culture or the transformation of normal cells into cancer cells?

We showed that when mixed with DNA from mammalian tissues, carcinogens induce conformational changes in native DNA from cancer tissues [*Beljanski* et al., 1981a] and in destabilized DNA from bacterial cells [*Beljan-*

ski et al., 1982a, b; *Le Goff and Beljanski,* 1981] with a much more pronounced effect than in normal DNA. The incidence of these phenomena on replication and transcription was hardly studied at all. For this reason we have assembled the main observations on the influence of carcinogenic, neoplastic and other compounds, in order to discover a common and coherent link between them. This will then be developed into a general model in chapter 6.

Interaction of Carcinogens with Chromatin

In normal and cancer cells [*Allfrey* et al., 1978; *Thurnherr* et al., 1973; *Ginsburg* et al., 1973], numerous studies have been carried out to determine the effect of carcinogenic agents, particularly on cell metabolism and on the destabilization of specific proteins, histone proteins in particular. Thus, for example, in the nuclei of the epithelial adenocarcinoma of rodents (induced with 1,2-dimethylhydrazine), two proteins (TPN_1 and TPN_2) are synthesized and become the predominant nuclear proteins of tumor cells. After treatment with DNase I, which acts on DNA segments to be transcribed into mRNA, chromatin loses the TPN_1 protein. It appears that DNase acts more efficiently on nucleosomes rich in acetylated histones (essentially H_3 and H_4), because acetylation appears to alter the chromatin configuration at the level of nucleosomes. This indicates that the most acetylated histones are localized in the activated regions of chromatin for transcription. However, the role played by TNP_1 and TNP_2 proteins is not known. Although not detectable in normal cells, they are very abundant in cancer cells [*Allfrey* et al., 1978; *Hawks* et al., 1974]. Available data seem to suggest that in the nuclei of cells where proteins are not expressed, corresponding genes where not degraded by DNase I. This shows that to be replicated or transcribed, DNA has to be separated from histone proteins. Acetylation of proteins in nucleosomes by butyrate furthers the synthesis of RNA in cancer cells (HeLa) to a greater extent than in control cells which did not undergo acetylation [*Ginsburg* et al., 1973]. These results suggest that acetylation of histones H_3 and H_4 destabilizes the DNA-histone complex, making DNA more susceptible toward DNase I. (Spacer DNA is selectively sensitive to another DNase, originating in staphylococci.)

Does a carcinogen act in situ directly on DNA in nucleosomes containing H_3 and H_4 histones before or only after they have been acetylated? There is no clear answer to this question, but the fact remains that carcinogens do induce the expression of genes.

In HeLa cells (clone 65 and 71), butyrate, hydroxyurea, methotrexate and Ara-C [*Ghosh and Cox*, 1976, 1977; *Ghosh* et al., 1977] might inhibit the synthesis of DNA without interfering with that of RNA. The amount of certain peptides and hormones produced by HeLa cells in the presence of the substances cited is increased by about 300%. This observation and the appearance of TNP_1 and TNP_2 proteins induced by 1,2-dimethylhydrazine account for gene activation. Ethionine, a powerful carcinogen, also induces activation of embryonic genes [*Hancock* et al., 1976] and elevates progesterone levels [*Sharma and Borek*, 1977]. However, the molecular mechanism for this gene activation has not yet been defined.

During the release of new information or when a simple amplification of preexistent information is expressed, modification of chromatin organization could lead to conformational changes, resulting in an acceleration of gene expression. This is a phenomenon in which the synergist interaction of endogenous and exogenous molecules on nucleosomes may be direct or may take place through histones or other DNA-related proteins such as untwisting and unwinding proteins. Carcinogenic agents or other compounds cited in this chapter allow the synthesis of RNAs and proteins mostly in cancer cells which are more susceptible than normal cells. The vulnerability of DNA in chromatin, besides conformational modifications, may be due to the acetylation of histones to different degrees and particularly to the capacity of DNA to undergo strand separation in cancer cells. These causes are closely interdependent.

Therapeutic Agents and DNA Release from Chromatin

Apart from substances well known as carcinogens and classified as such, there exist numerous antimitotic agents and even anticancer drugs which have been revealed as acting on in vitro cell differentiation [*Lotem and Sachs*, 1979; *Friend* et al., 1971], on in vitro cancer DNA synthesis [*Beljanski*, 1979], and even as carcinogens in vivo when used over a long period [*Kawamata* et al., 1958; *Svoboda* et al., 1970; *Haidle*, 1971]. Since the effect of these substances, when compared to classical carcinogens, is slight, they enable one to get a clear understanding of the mechanism which induces changes in the cell. These compounds separate the strands of spacer DNA which in turn becomes more susceptible to specific nucleases: these may attack the nucleosomes themselves [*Lotem and Sachs,* 1979], in order to eliminate certain histone proteins, thus destabilizing their structure. This process might be followed by DNA strand separation. Thus adriamycin [*Dimarco,* 1975], which is an antibiotic but also a carcinogen [*Bertazzoni* et al., 1971], when incubated with nuclei from fibroblasts (BALB/3T3 line) in the presence of ^{3}H-thymidine and nuclease (of *Neurospora crassa*), induces local DNA strand separation in the nucleus. This enables the nuclease to degrade the liberated strands. The nuclease used is highly specific for the degradation of single-stranded DNA [*McGuire* et al., 1976]. In contrast, when nuclei are incubated in the presence of adriamycin alone, without the nuclease [*Center,* 1979], no change in the chromatin structure is observed. Thus, in the presence of this nuclease, daunorubicin and ethidium bromide also make DNA susceptible to this enzyme. DNA strand liberation in the presence of the nuclease may be assayed either by the appearance of acid-soluble nucleotides in the incubation mixture or by measuring the integrity of DNA on a sucrose gradient after ultracentrifugation. This phenomenon does not occur with native DNA isolated from nuclei, and this suggests that in the nucleus certain molecules contribute to DNA strand separation in cooperation with the above mentioned agents. Actinomycin D (20–100 μg/ml) appears to be inactive with nuclear DNA incubated in the presence of *Neurospora crassa* nuclease. The choice of DNA as well as the concentrations are crucial when observing the effect of adriamycin, daunorubicin and ethidium bromide.

Considerable efforts have been made to demonstrate the effects of neoplastic substances and carcinogens on normal cells, on their DNA, and on the steps involved in the transformation of cancer cells into normal cells. However, the in vitro behavior of DNA from normal and malignant tissues has been neglected, although such a comparison is the key for determining the molecular mechanism of gene activation.

In fact, it is possible to compare the biological activity of template DNA isolated from healthy and cancerous cells on in vitro DNA synthesis catalyzed by DNA-dependent DNA polymerase I. The results of such a comparison show that DNA templates do not allow synthesis with the same efficiency, particularly when the reaction mixture contains carcinogens or an antimitotic [*Beljanski,* 1979; *Beljanski* et al., 1981a]. This observation, and the physicochemical difference between many cancer DNAs and the corresponding normal DNAs, provide an excellent basis for an understanding of the release mechanism of specific information or gene activation.

Destabilization and Condensation of Chromatin

Acetylation and phosphorylation of histones, methylation of RNA and proteins and formation of poly (ADP-riboxyl-)polymers are guide events which allow the study of changes involved in chromatin condensation or disorganization. These changes appear to be necessary for the progression of cellular cycles. Phosphorylated H_1 histone seems to be involved in the initiation process for chromosome condensation. Zn chloride, which inhibits in vivo phosphatase activity and, consequently, maintains H_1 histone as phosphorylated protein in metaphase, does not prevent the separation of chromosomes at the end of mitosis. Phosphorylation appears to control the *initiation* of mitosis: a low level of phosphorylation of H_1 histone leads to incomplete condensation of chromatin [*Matsumoto* et al., 1980]. Does Zn bind to histones or to DNA, or does it change the activity of the enzyme, regulating the maintenance of the chromatin structure? Using rat liver microsomes in an in vitro system to cause metabolic activation of ^{3}H-benzo(a)pyrene (BP), the amount of carcinogen that binds to DNA from calf thymus nuclei is of the order of 1 per 10^4 bases [*Jahn and Litman,* 1978]. However, non-covalent fixation of BP also takes place with DNA, RNA and even with proteins. The fact that the totality of BP is detached from DNA after exogenously added staphylococcal DNase action suggests that the compound was fixed essentially on 'spacer DNA'. This DNA segment links nucleosome cores. It has been shown that BP and ethidium bromide display a strong preferential binding to DNA associated with H_1 histone which seems to be in contact with spacer DNA [*Jahn and Litman,* 1978]. Thus, these two compounds may intercalate between DNA strands [*Le Pecq,* 1978]. Staphylococcal DNase releases 40–60% of radioactive BP. This enzyme acts on internucleosomal DNA. DNase I, which acts on 'exon' DNA, representing activated genes, liberates only 8–12% of BP.

It will be important to discover whether other carcinogens or agents bind on spacer DNA or on nucleosome cores and whether their mode of action is to disorganize these latter structures which contain functional genes [*Allfrey* et al., 1978].

Interaction of Carcinogens and Neoplastic Agents with DNA

Carcinogens may be attached to DNA by low physiological interactions [*Ryser,* 1971], hydrogen bonds [*Sobell,* 1973; *Sarma* et al., 1975; *Nagata* et al.,

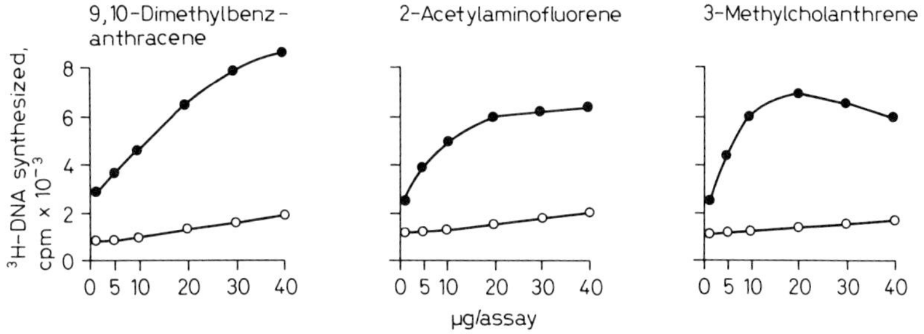

Fig. 15. Effect of carcinogens on DNA synthesis in vitro. ● = DNA from breast cancer tissue; ○ = DNA from breast tissue [from *Beljanski*, 1979].

1966], covalent linkage [*Meehan* et al., 1976; *Warwick*, 1967] or even by electrostatic forces [*Allfrey* et al., 1978]. Some of these agents require activation to convert carcinogenic compounds into electrophil intermediaries which then transfer their methyl (or, in the case of alkylating agents, ethyl) groups to DNA, in which guanine is the main acceptor. They also interact with nucleophilic substances or – to be more precise – with nucleophilic centers occurring in nucleic acids, but also with proteins, amino acids or lipids. Quite a few carcinogens and mutagens do require activation. There exist in vitro systems [*Beljanski*, 1979; *Beljanski* et al., 1981a, 1982b] in which, without activation, all carcinogens react with soluble DNA. This would suggest that activation of carcinogens is possibly necessary for their penetration into cells but not obligatory for their direct effect on DNA. The Oncotest [*Beljanski* et al., 1981a] has shown that carcinogens which apparently require activation in the *Salmonella*/microsome test react directly and efficiently, particularly with cancer DNA, in the absence of activating enzymes. They strongly stimulate cancer DNA synthesis in vitro but only slightly that of DNA from healthy cells (fig. 15). Identical results have been obtained using purified DNA from tester strains of *Salmonella typhimurium* which contain destabilized chains [*Beljanski* et al., 1982a].

The base composition of DNA from animals and plants varies [*Shugalin* et al., 1970], but only within narrow limits. The ratio of (G-C)/(T-A) nucleotides is characteristic for DNA of a given species [*Woese*, 1967]. The asymmetry of DNA chains, some of which contain excess pyrimidine residues [*Mazin*, 1976; *Strauss and Birnboin*, 1976], and the presence of poly-d(A) – poly-d(T) segments in DNA isolated from various sources [*Jacobson* et al., 1974; *Shenkin and Burdon*, 1974; *Mol* et al., 1976] illustrate the heterogeneity of nucleotide sequences in different DNAs. This offers a possibility for differential interaction between DNA and various substances, including carcinogens and neoplastic agents. Thus, olivomycin selectively binds to G and C bases in DNA, and this accounts for chromosomal fluorescence [*Van de Sande* et al., 1977]. Hoechst 33258 antibiotic preferentially binds to A-T sequences [*Latt and Wohlleb,* 1975]. In order to allow interaction, DNA strands must be accessible to reactive agents and should not be engaged in a complex chromatin structure. The main target of these agents is DNA or, to be more precise, 7-nitrogen or 6-oxygen of guanine [*Connors,* 1975; *Ludlum,* 1975]. The 7-nitrogen of guanine itself may bind 90% of an alkylating agent. Depending on concentrations of alkylating agents, guanine might be alkylated on only one of the two DNA strands. This causes physicochemical repercussion in the structure of DNA [*Iyer and Szybalski,* 1963; *Laval,* 1977]. Alkylated segments of DNA can be repaired by cooperative action of endonuclease, DNA polymerase, polynucleotide ligase or by the so-called DNA-repairing SOS system [*Witkin,* 1977; *Prakash and Strauss,* 1970] described for bacteria. Even lesions as strong as 'intercrossing' linkages (by substances which intercalate and form a bridge between DNA strands) are only repaired slowly [*Cole* et al., 1976; *Fujawara* et al., 1977]. Alkylation might also occur with adenine, cytosine or even with phosphates. In this latter case, alkylating agents produce phosphotriesters which favor the rupture of phosphodiester linkages in DNA [*Singer,* 1977]. Such interactions provoke physicochemical changes in DNA. The compound 1, 3-bis(2′-chloroethyl)-*l*-nitrosourea (BCNU), an antimitotic and carcinogenic [*Tashima* et al., 1979; *Wilson* et al., 1970] but non-alkylating agent, interacts with DNA from L 1210 leukemic cells of mice and modifies the helical structure of DNA. This results in the inhibition of DNA chain elongation [*Tashima* et al., 1979], leading to an accumulation of relatively short segments of DNA. It is not clear if this agent acts in a similar way in normal cells [*Iriarte* et al., 1966] and brain tumors [*Wilson* et al., 1970]. It probably acts by separating double strands of the DNA in tumor cells.

Intragenic distribution of a carcinogen in DNA is poorly understood because it is difficult to determine its localization, due to the extreme complexity of the DNA molecule. Depending upon doses used, a carcinogen, for example DMBA, binds with preference to repetitive sequences of DNA rather than to DNA of the homopolymer type [*Shoyab,* 1978]. In fact, at low concentrations metabolites of this carcinogen bind preferentially to certain segments of DNA while at high concentrations they are equally located on all repetitive bases of DNA, as was demonstrated by the DNA-DNA hybridization technique [*Shoyab*, 1978]. A given carcinogen may in vivo lead to different biological effects [*Britten and Davidson*, 1969]. Depending upon doses used, the carcinogen exhibits a stimulatory or inhibitory effect on cancer cells. This has been demonstrated in vitro and in vivo.

Intercalating Substances and DNA

Another family of substances, called intercalating agents, inhibit directly the synthesis of DNA by establishing solid links between two base pairs on opposite strands of the double helix, thus inducing modification in the DNA structure [*Lerman*, 1961; *Waring*, 1965; *Wang*, 1971]. These conformational modifications allow interactions with different enzymes, histones and other proteins. Thus, the 'DNA-actinomycin S' complex [*Sobell*, 1973] has consequences on the activity of RNA polymerase. Transcription of DNA is consequently inhibited [*Schwartz*, 1974].

Daunorubicin and adriamycin, which intercalate between DNA strands, may 'unwind' DNA [*Waring*, 1970] as well as open DNA strands [*Beljanski* et al., 1981a]; in this way, the template capacity of DNA is increased for replication and transcription.

Absorbance changes and helix stability of DNA after fixation either of daunorubicin (or its derivatives) or of adriamycin have been studied [*Johnston* et al., 1978; *Plumbridge and Brown*, 1978]. Daunorubicin does not bind to particular bases in DNA [*Latt and Wohlleb*, 1975]. The glucopeptide bleomycin, which is both an anticancer agent [*Umezawa*, 1976] and a carcinogen [*Haidle*, 1971; *Beljanski*, 1979], binds to the double DNA structure or to a single strand in the presence of ferrous ions, inducing sequence cleavages on G-T and G-C base pairs [*Müller*, 1978; *Müller* et al., 1972]. However, a strong concentration of ferrous ions [$Fe(NH_4)SO_4H_2$] may by itself cause a rupture of nonspecific sites [*Takeshita* et al., 1978; *Le Pecq*, 1978; *Umezawa*, 1976]. One bleomycin molecule binds by noncovalent linkage to about 350 nucleotides [*Müller*, 1978; *Müller* et al., 1972], and this reaction is specific for DNA strands. RNA is not degraded by bleomycin. In a DNA-RNA hybrid, only the DNA strand is depolymerized by high bleomycin concentrations [*Haidle and Bearden*, 1965]. Nicked DNA may be repaired in vivo by DNA-repairing enzymes, as already stated [*Byfield* et al., 1976; *Cole* et al., 1976; *Fujawara* et al., 1977]. Actinomycin D, which intercalates between DNA strands, is generally used as an inhibitor of transcription, but in very low concentrations it leads to very surprising and important results. For example, at doses of 0.03–0.10 μg/ml, actinomycin D causes significant stimulation of the axon formation in neuroblastoma cells of mice (C 1300) and in cells cultured in vitro [*Bear and Schneider,* 1979]. This effect can be observed if cells are incubated over a 20-hour period. Beyond this, however, there is a diminution of axon formation. In certain cases actinomycin D may also stimulate protein biosynthesis [*Leinwand and Ruddle,* 1977; *Palmiter and Schimke,* 1973; *Craig,* 1973; *Goldstein and Penman,* 1973]. Stimulation of the axon formation is dose-dependent, and at pH 6.6 and 7.4 it is amplified 2.7- and 7-fold, respectively. (It should be recalled that alkaline pH destabilizes DNA while low pH stabilizes it.) In contrast, high concentrations of carcinogens cause general toxicity rather than inhibition of macromolecular synthesis [*Bear and Schneider,* 1979]. Actinomycin D (1 ng/ml) induces differentiation of myeloid leukemic cells (HL60) into granulocytes. DMSO and 12-O-tetradecanoyl-phorbol-13-acetate (TPA) induces differentiation of these same cells into macrophages [*Lotem and Sachs,* 1979]. Thus, under the influence of one or another of these substances, human leukemic cells release specific informations, leading to differentiated cells. Actinomycin D is involved also in the dedifferentiation of normal cells in the presence of carcinogens. The carcinogen DMBA causes initiation of skin tumorigenesis which is first accompanied by RNA synthesis. This is inhibited by actinomycin D [*Reich* et al., 1961; *Gelboin and Klein,* 1964]. These results might be explained in the light of our own results which show that DMBA, like actinomycin D, binds to DNA and, depending on the dose, succeeds or fails in making certain segments of DNA functional. The action of such substances may well be either synergistic or antagonistic [*Beljanski* et al., 1981] (chapt. 6).

Data available indicate that sequences of DNA from normal and cancer cells are identical [*Le Pecq,* 1978; *Sandberg and Sakurai,* 1974], and some authors conclude that 'it is almost certain that toxicity of these compounds (carcinogens) cannot result in a preferential reaction of alkylating agents with DNA of cancerous cells' [*Le Pecq,* 1978]. However, a great many facts contradict these assertions. Potentially carcinogenic substances do induce differentiation of cancer cells into normal types, while few or even no examples have been reported of a transdifferentiation taking place in normal cells. We have shown that DNAs isolated from normal and cancer tissues behave differently when they are incubated in vitro in the absence or presence of carcinogens. The carcinogens preferentially and strongly stimulate cancer DNA synthesis, but hardly at all that of the normal DNA [*Beljanski,* 1979; *Beljanski* et al., 1981a]. This difference is also confirmed by the observation that 'specific anticancer agents' inhibit the in vitro synthesis of all cancer DNAs tested but normal DNA synthesis is only slightly or not at all affected by them [*Beljanski and Beljanski*, 1982; *Beljanski* et al., 1982b]. These substances are lethal for cancer cells, but the same doses are not toxic for normal cells, cultured in vitro or carried by mice. These facts may be explained by the different reactivity of cancer and normal DNAs whose hyperchromicity is modulated in the presence of these various substances [*Beljanski*, 1979; *Beljanski* et al., 1981a]; (see chapt. 6). In addition, we have recently found that DNA, isolated from wild-type (His^+) *Salmonella typhimurium*, reacts in a totally different way in the presence of carcinogens than DNA from the mutant His^- [*Beljanski* et al., 1982a]. This certainly accounts for the results of the mutatest [*Benedict* et al., 1977]. It is now clearly established that DNAs, depending upon the physiological state of the cell from which they are extracted, do not in all cases exhibit the same affinity toward carcinogens. Since chemical carcinogens stimulate cancer DNA synthesis in vitro more efficiently than that of normal native DNA, it may be concluded that the former contains more single-stranded segments than the latter and these segments may be preferential targets for carcinogens.

Release of Specific Information and DNA Replication

Experimental data suggest that fresh transcription can be performed only if a new replication has first been carried out. Lymphoblasts, for example, generate in vitro some characteristic lymphocytes only if mitosis has first been stimulated by phytohemagglutinin [*Resch* et al., 1977]. In contrast, colchicine and vinblastine (10^{-4} and 10^{-5} *M*, respectively), added to a lymphocyte culture which was pretreated with concanavalin A, prevent DNA synthesis but do not affect the synthesis either of RNA or specific protein (lymphotoxin). Release of information takes place in spite of the inhibition of DNA synthesis. These results may be explained in the light of our recent studies which demonstrated the capacity of the above substances to open DNA double strands in a particular region [*Beljanski* et al., 1981a], thus permitting synthesis of specific mRNA. Let us now compare these results with those showing that 3-methylcholanthrene, a carcinogen, provokes an increase in the amount of RNA synthesized in liver cell nuclei and an increase of mRNA activity in a normal and active gene system [*Loeb and Gelboin*, 1963]. The normal gene system denotes the whole series of biochemical processes which lead from a gene to the phenotypic character by which it is recognized [*Waddington,* 1962]. It seems fairly certain that carcinogens, in attacking DNA close to initiation sites, prevent DNA synthesis while RNA synthesis, whose initiation sites are different, is not inhibited.

Colchicine, vinblastine, insulin or hydrocortisone, in association with prostaglandin F2, strongly increase the initiation rate of DNA synthesis, provided that quiescent and confluent cells (Swiss 3T3) were already aroused by prostaglandin [*Baserga,* 1976].

Some results were obtained when prostaglandin was replaced by fibroblast growth factor [*Gospodarowicz,* 1974]. No clear-cut interpretation of the effect of these different agents was proposed. However, it was suggested that colchicine might act on chromosomes, facilitating transcription of certain specific mRNA [*Otto* et al., 1979]. The observation that, in association with prostaglandin F2, agents significantly increase the initiation rate of DNA synthesis lends weight to our interpretation that the effect is probably on DNA strand separation, favoring the access of DNA-dependent DNA polymerase and consequently the increase of the amount of synthesized DNA [*Beljanski* et al., 1981a]. A good correlation between increased DNA strand separation and increased DNA template activity (transcription) strengthens this view.

These steps might be different in the presence of the same substance. We have already emphasized that DMSO concentrations play an important role in these processes [*Friend* et al., 1971; *Tanaka* et al., 1975]. Friend leukemia cells grown with 2% DMSO undergo changes similar to those observed in normal erythroid differentiation [*Harrison,* 1976]. We interpret these results as follows: genes for hemoglobin synthesis inactive in leukemia cells become active because DMSO allows DNA strand separation and the release of specific information. In numerous studies, DMSO was used as a solvent to dissolve compounds which were not soluble in water [*Sarasin and Moulé,* 1976; *Levine,* 1975; *Benedict* et al., 1977]. DMSO also facilitates the penetration of drugs or carcinogens into rat liver cells. A single DMSO injection provokes duplication of endoplasmic reticulum [*Gal* et al., 1978]. In addition to the observation that DMSO opens DNA double strands [*Beljanski* et al., 1981a], it might also act to change the activity of certain enzymes, RNase for example. In fact, reticulum fractions of animals treated with DMSO contain less RNase activity than fractions from control animals. It would appear, therefore, that mRNA free or bound to ribosomes from treated animals is less damaged by RNase and that this results in increased protein synthesis. Less RNase activity in the presence of DMSO could simply be due to dissociation of the enzyme-substrate complex. It should be noted that in vitro incorporation of amino acids is influenced by the activity of RNases bound to ribosomes [*De Groot* et al., 1976]. When used at non-inhibitory concentrations for growth, DMSO induces differentiation of a leukemic cell line, namely murine virus leukemic cells [*Collins* et al., 1978]. In a medium containing 0.5–1.0% of DMSO, cells multiply with practically the same speed as control cells. Differentiation of erythroid cells is accompanied by active hemoglobin synthesis. It was postulated that DMSO might stimulate the synthesis of nucleic acids and proteins [*Hellman* et al., 1967; *Hagemann,* 1969]. In addition, it appears to affect the secondary and/or tertiary structure of macromolecules [*Katz and Penman,* 1966; *Rammler and Zaffaroni,* 1967; *Strauss* et al., 1968]. Axon overgrowth in neuroblastoma cells takes place even when the serum was removed from the culture medium [*Seeds* et al., 1970], and differentiation is induced in slime molds by removal of food to the plasmodium [*Rusch,* 1970]. These results suggest that differentiation takes place in the absence of growth and that the synthesis of DNA is probably slowed down to allow expression of specific genes through mRNA synthesis. DMSO does not always induce the events in a program to completion. Induction of globin mRNA in erythroleukemic cells has been studied by treating the cells either with DMSO or with heme, under conditions permitting viability of normal cells [*Lowenhaupt and Lingrel,* 1979]. In the presence of each of these two agents, a rapid 4-fold accumulation of globin mRNA was observed. In cells treated with DMSO, however, a globin mRNA is affected by stability changes, and the capacity of cells to proliferate is modified. In contrast, heme does not induce such changes. Dividing cells accumulate globin mRNA, and this accumulation, although limited, may continue even when cells do not divide. The mRNA synthesis and stability constitute elements included in the differentiation process which involves either activation or inactivation of RNases. Differentiation of erythroleukemic cells induced by DMSO is accompanied by a parallel change in the level of enzymes involved in purine metabolism [*Gusella and Housman,* 1976]. However, it has been shown that hypoxanthine, thioguanine and mercaptopurine, which, as free bases, induce differentiation of erythroleukemic cells, interact directly with a discrete cellular target whose nature needs to be identified and characterized in order to elucidate the control mechanism of the differentiation process. DMSO acts by inducing large amounts of actin in all nuclei and as soon as DMSO is eliminated, the phenomenon disappears. The maximum effect of DMSO is manifested at concentrations ranging from 5 to 10%, but the phenomenon does not appear when cells are incubated in the presence of glycerol which prevents DMSO penetration into cells. Mg^{2+} inhibits the DMSO effect. It is well known that Mg^{2+} strongly stabilizes DNA double strands [*Fukui and Katsuma,* 1980]. DMSO and dimethylformamide induce terminal differentiation of human promyelocytic leukemic cells, but the mechanism of this induction is not known [*Collins* et al., 1978]. Both agents induce also electrophysical changes in a line of mouse neuroblastoma cells [*Kimchi* et al., 1976], differentiation of erythroid cells in rat leukemia [induced by DMBA: *Kluge* et al., 1976], and also lysozyme production in leukemia cells [*Krystosek and Sachs,* 1976]. Differentiation or inactivation of protein synthesis does not take place in the presence of certain DMSO concentrations, and numerous experiments have been carried out with excessive concentrations in order to observe some interesting phenomena. Among the interpretations put forward to explain the action of DMSO, we should keep in mind the suggestion that DMSO diffuses into cell nuclei, changes conformation of the DNA-chromatin complex, and thus permits initiation of transcription of gene(s) coding for specific differentiation processes [*Tanaka* and al., 1975]. It is rather astonishing that DMSO, so often used as a solvent in the Ames test for carcinogenesis, provokes differentiation of certain cells and thus might considerably modify conclusions based on the assumption that mutations induced by carcinogens become activated in the presence of DMSO. Many examples prove that DMSO behaves as a biological substance, permitting the release or expression of specific information. Thus, preincubation of Novikoff hepatoma cells in the presence of (2.5%) DMSO results in a 400% increase of thymidine incorporation into DNA [*Barra* et al., 1978]. When added in low concentration to the incubation medium, histones do not contribute to the DMSO action. This is not surprising since DMSO has the property of dissociating the DNA-protein complex. Here again, DMSO concentration is of major importance. These phenomena are no longer observable at a concentration of 20% or more, probably because of the arrest of cellular synthesis. It was recently demonstrated that at a given concentration DMSO irreversibly inhibits the fusion of myoblasts, giving rise to multinuclear myotubes [*Blau and Epstein,* 1979]. After DMSO elimination, L 8 cells cease to synthesize muscle protein, and several days after the cells have achieved the state of confluence, almost all cells continue to synthesize DNA in the presence of DMSO. However, such cells are not capable of entering into the G_0 phase of the mitotic cycle necessary for the initiation of cell differentiation. By enforcing thus a continuous proliferation of cells, DMSO inhibits all parameters required for differentiation in a coordinated manner. Once cell fusion and synthesis of specific muscle proteins are in progress, addition of DMSO does not further modify events. As soon as differentiation starts, DNA synthesis in control cells is stopped or slowed down in the region where multiplying cells are in fusion. It should be emphasized that the capacity of DMSO to maintain muscle cells in an active proliferating phase prevents differentiation [*Cohen* et al., 1977; *Holtzer* et al., 1973].

However, when DMSO causes differentiation of erythroleukemic cells, its presence appears necessary for DNA synthesis and probably somewhat later in order to facilitate differentiation of cells [*Levy* et al., 1975].

These results differ from those showing that DMSO prevents differentiation of myeloblastic cells [*Blau and Epstein,* 1979]. In fact, these two biological systems behave differently. A change in the conformation of DNA, or the DNA-protein complex, may be produced in the presence of DMSO and thus initiate transcription of genes which regulate the expression of information in erythroid cells [*Tanaka* et al., 1975]. It should not be forgotten that DMSO has also teratogenic effects [*David,* 1972; *Caujolles* et al., 1967] and behaves like a carcinogen [*Beljanski,* 1979]. Differentiation of Friend erythroleukemic cells by DMSO shows in addition that it acts in this system on a site where viral mRNA transcription takes place, or at least that it inhibits degradation of viral mRNA. In no case does it act as an inducer in the release of viruses [*Colletta* et al., 1979]. Cells which have differentiated during release of information under the influence of DMSO maintain their properties. New nucleotide sequences are expressed as measured by the increase in the percentage of RNA/DNA hybridization at the saturation level. This does not happen in control cells. Cells simultaneously treated with DMSO and BrdU synthesize more mRNA and specific polypeptides of virus, but differentiation does not take place. These results clearly show that DMSO has a particular biological role in the release of information due to its aptitude to induce DNA strand separation with greater efficiency in cancer cells, irrespective of their origin [*Beljanski* et al., 1981a].

Several conclusions might be drawn from experiments performed with DMSO. These are:

(a) DMSO acts on the permeability of cells [*Friend and Freedman,* 1978], diffuses into cell nuclei and alters the conformation of the DNA-chromatin complex.

(b) The effect of DMSO on those cells which have not yet undergone differentiation is reversible.

(c) When used at concentrations which do not inhibit growth, DMSO induces in vitro differentiation of different malignant cell lines. By imposing continuous proliferation on normal myoblasts, it reversibly prevents the formation of multinuclear myotubes.

(d) DMSO, being a bipolar and alkylating molecule, permits release of specific information due to the separation of DNA double strands and thus provides single-stranded DNA segments [*Beljanski* et al., 1981a].

Phorbol Derivatives and Induction of in vitro Cell Differentiation

Beside carcinogens, antimitotics and steroids, it has been observed that phorbol derivatives, considered by some authors as promoters of carcinogenesis [*Hennings and Boutwell,* 1970; *Berenblum,* 1941] and by others as its initiators [*Hecker,* 1968; *Kopelovich* et al., 1979; *Kopelovich,* 1982], may induce differentiation of leukemic cells into different types [*Lotem and Sachs,* 1979; *Huberman and Callaham,* 1979]. In other words, in the presence of appropriate concentrations, they may behave like DMSO or like the substances already mentioned.

The differentiation of human promyelocytic leukemic cells in vitro (HL 60) in the presence of TPA has been studied [*Huberman and Callahan,* 1979]. The percentage increase in myelocytes, metamyelocytes and other myeloid cells, as well as in cells for phagocytosis, has been determined. At 6×10^{-11} *M,* TPA induces terminal differentiation. At much higher concentrations, however, it provokes cell degeneration. In this same cell line, it induces differentiation of promyelocytic cells into macrophages [*Lotem and Sachs,* 1979], while neither lipopolysaccharide (10 μg/ml) nor phytohemagglutinin (0.02 mg/ml) or dexamethasone (400 μg/ml) induces this differentiation. When applied to normal bone marrow cells, TPA (0.05 μg/ml) stimulates the activity of a protein acting as an inducer of cell differentiation into macrophages or granulocytes and sensitizes the myoblasts to synthesize this protein. Recent studies [*Kasukabe* et al., 1981] have shown that induction of differentiation of mouse myeloid cells depends not only on the presence of TPA but also on the type of serum present in the medium. It should be stressed that the serum contains RNase which may provide RNA primers for DNA replication once DNA strands have been locally separated by TPA.

In other systems, at relatively high doses, phorbol monoacetate (PMA), 10^{-4}–10^{-5}*M,* inhibits spontaneous or induced differentiation of chicken myoblasts [*Cohen* et al., 1977; *Ishi* et al., 1978] and erythroleukemic cells in mice [*Rovera* et al., 1977]. It would be interesting to know whether or not very low PMA concentrations induce differentiation in these cells. Perhaps a certain specificity between the DNA of a chosen cell type and PMA could be discovered.

It has been suggested that tumor promoters act by inducing mitotic recombination which could lead to the expression of recessive carcinogen-induced mutations and the transformation of normal cells into tumor cells [*Kinsella and Radman,* 1978]. More data are needed to consider the possibility that tumor promotion by TPA results primarily from enhancement of mitotic recombination. TPA alone induces in vitro the transformation of normal and cancerous cells and stimulates the growth of human fibroblasts which have been genetically predisposed for cancer [*Kopelovich* et al., 1979]. In other cases, normal cells treated with tumor promoters temporarily exhibit the properties of transformed cells [*Weinstein* et al., 1978]. The powerful tumor promoter TPA also provokes stimulation of nucleic acid synthesis as well as numerous phenotypic changes in different cells [*Raick,* 1973; *Sivak and Van Duuren,* 1970:; *Hennings and Boutwell,* 1970]. The stimulated incorporation of radioactive precursors into mice skin RNA by croton oil, containing phorbol esters [*Hecker,* 1968], is of great interest. The amount of synthesized rRNA is proportional to that of the phorbol ester used [*Baird* et al., 1971], and one single application of this ester to mice skin provokes a significant synthesis both of rRNA and ribosomes [*De Young* et al., 1977]. Recent investigations have produced some first-rate examples of reactivation by TPA of silent genes for rRNA [*Soprano and Baserga,* 1980]. In hybrid cells, obtained by fusion of fibrosarcoma cells (HT 1080) with macrophages from BALB/c mice, genes for rRNA of both cells are present. Cells synthesize human 28S rRNA but not mouse 28S rRNA [*Perry* et al., 1979]. After a 24-hour treatment by TPA of a hybrid cell clone (55-54), both human and murine 28S rRNA are synthesized, although the expression of mouse rRNA is independent of the TPA concentration. The dose-response curves are not perfect, probably due to TPA instability. Certain authors claim that the mode of action of TPA in gene reactivation is obscure, but we maintain that it may be explained by the action of phorbol derivatives on DNA strand separation which is a qualitative and quantitative 'action' of newly liberated information [*Beljanski* et al., 1981a]. Those phorbol derivatives, which do not promote tumor appearance (phorbol 13,20-diacetate, for example), do not activate genes for 28S rRNA in hybrid cells [*Perry* et al., 1979] and do not induce DNA strand separation [*Beljanski and Le Goff,* in press]. TPA can also exert membrane and cellular effects on the function of the epidermal growth factor (EGF) receptors and thus might be of assistance in elucidating certain aspects of cell membrane structure and function related to growth control and gene expression [*Lee and Weinstein,* 1979].

A tumor promoter in two-step carcinogenesis (TPA) induces within 12 h after a single application on the dorsal skin of mice a nearly 10-fold increase of histidine decarboxylase. The activity of ornithine decarboxylase is also increased, but significantly earlier than that of histidine decarboxylase. No mechanism of these activities after a single application of TPA has been proposed [*Watanabe* et al., 1981]. TPA induces transglutaminase activity in epidermal basal cells, cultured in vitro. This enzyme is responsible for the formation of the cross-linked envelope in differentiated cells [*Yuspa* et al., 1980].

There is a striking and obvious parallelism between gene expression and the capacity of these substances to separate the double strands of DNA, needed for gene expression. Phorbol derivatives, DMSO, PMA, carcinogens, antimitotics and steroids, all inducing gene expression, have in common the capacity to open the double strands of DNA. This cannot be fortuitous.

PMA, isolated from croton oil, is a very active promoter of tumor cell growth in animals but only if administered either just before or just after intraperitoneal injections of tumor cells [*Berenblum,* 1975]. In the presence of chick embryo fibroblasts, PMA ($5 \times 10^{-9} M$) induces an increase in the amount of plasminogen activator, and this produces a whole series of illustration indicating phenotypic cellular changes after PMA influence [*Wigler and Weinstein,* 1976]. These changes disappear as soon as PMA is removed from the medium. This suggests that PMA is not carcinogenic in nature.

PMA induces DNA strand separation [*Beljanski* et al., 1981] and, depending on its concentration, this might be local corresponding to the gene responsible for the synthesis of the plasminogen activator or some other related proteins. Morphological cell changes caused by phorbol esters are accompanied by the appearance of dense cellular aggregates [*Quigley,* 1979]. The importance of certain proteins in this phenomenon is confirmed by protease inhibitors which, by preventing the activity of these latter, prevent PMA-induced morphological alteration. Inhibition of proteases may result in the absence of a sufficient quantity of certain peptides which, by fixing to DNA, contribute to DNA strand separation by PMA. A number of particular peptides isolated from cancer cells significantly inhibit DNA transcription for globin mRNA and RNA translation in reconstituted cell-free, peptide-free systems. These peptides appear to stabilize the double-stranded DNA molecule, thus providing an endogenous control of gene expression. It was suggested that peptides may control the length of an active genetic unit (gene) and that in carcinogenesis the level of peptides is strongly reduced for specific genes by the action of induced peptidases [*Hillar and Przyjenski,* 1973]. It should be noted that induction of proteases is one of the effects of tumor promoter action [*Blomberg and Robbins,* 1975] and that inhibitors of proteases prevent tumorigenesis in mouse skin [*Troll* et al., 1970].

Phorbol-12,13-dimyristate (PDM), introduced in vitro into a culture medium of epidermic cells, induces hyperplasia [*Slaga* et al., 1976]. This could be due to an increase in DNA, RNA and protein synthesis, involving no cell division. TPA stimulates DNA synthesis in cells maintained in culture in vitro [*Yuspa* et al., 1980], but its activity is completely dependent on the presence of serum in the culture medium. However, sera from various origins contain nuclease activity, RNase in particular, permitting formation of RNA fragments which are good primers for DNA synthesis [*Beljanski* et al., 1975], especially when DNA strands are separated. It was reported that serum could be replaced by insulin and a polypeptide which would act as an epidermal growth factor (EGF) [*Dicker and Rozengurt,* 1978]. We know that insulin acts as a modulator of RNase, of pancreatic RNase in particular [*Beljanski,* unpublished results]. Stimulation of DNA synthesis of human or mouse cells by low TPA concentrations and a growth factor might be due to substances which locally separate DNA strands, thus permitting RNA polymerase to function on several initiation sites. Given together but without serum, TPA and growth factor stimulate DNA synthesis of fibroblasts in a quiescent state [*Dicker and Rozengurt,* 1978].

Summary

Data described and analyzed in this chapter provide evidence that carcinogens, antimitotics, tumor promoters, DMSO and antibiotics, at given concentrations, permit the release of specific information in cells which are in most cases maintained in vitro. The appearance of specific mRNAs or proteins has been more frequently observed with cancer cells orginating from various sources than from normal cells. In these events, the concentration of agents used is decisive. At low doses, these substances induce in vitro cell differentiation while at high doses they provoke degeneration both of normal and of cancer cells. DNA, the ultimate target for these substances, contains sites which appear to be specific for a given agent which interacts with cellular DNA. From a given cell culture, different types of cells may develop according to the nature and concentration of the agents used. Although these agents exhibit their effect on various cellular macromolecules (RNA, DNA, proteins), in most cases the release of information uses a physicochemical mechanism which enables the agents, a carcinogen, for example, to cause differentiation. Under the influence of carcinogens, a correlation is established between cell multiplication in vivo, DNA synthesis in vitro, and DNA strand separation of given segments. The effects may be cumulative and additive in respect to the sites involved. Sequential effects on different levels of replication and transcription is the key mechanism for the activity and the future of the cell.

6. Basic Mechanism of Gene Activation

Introduction

Gene activation is considered to be an obligatory process in living cells. Classical notions of gene activation are based essentially on the observation of the appearance or amplification of one or several proteins in a cell, which endow this cell with new biochemical or morphological properties. The newly acquired state of a cell containing activated genes will depend on endogenous or exogenous agents which may persistently interfere with the maintenance of gene expression, or its repression.

Since genes are different in their chemical nature (nucleotide sequences) and since some of them are activated by a certain agent, while others are not, the agents involved in gene activation are obviously specific. This view is supported by observations that one substance or agent acting as gene activator induces, at given concentrations, the expression of one given gene whose product might affect the expression of other genes. Activation should proceed without altering the chemical structure of the genes. In contrast to physiological gene activation, gene expression might be altered by mutations which induce changes in DNA bases.

By itself DNA cannot initiate cell activity. Its response depends upon the surrounding substances. Activation of genes takes place under the action of various naturally occurring or chemically synthesized agents. In order to draw up a message, double-stranded segments of DNA corresponding to this message need to be opened at the points where transcription should occur. This happens under the action of activating molecules present in the cytoplasm or nucleus: enzymes, hormones, RNA, exogenous or pH-modifying agents. DNA can be visualized as a 'keyboard on which biological and chemical stimuli play' [*Grassé*, 1973].

Three essential situations should be considered:

(1) To maintain a cell which does not divide, numerous constituents need to be renewed. This requires sequential and differential RNA synthesis in concert with different enzymes which are directly or indirectly involved in

gene activation. In nondividing cells there is no need for complete DNA replication.

(2) DNA of dividing cells must be entirely and accurately replicated. Differential RNA synthesis must take place in such cells before they divide.

(3) Cell differentiation implies differential transcription of one or several genes as well as termination of the activity of some of them in order to change certain metabolic or morphological characteristics which will be transmitted to daughter cells. Partial or complete replication during differentiation is a matter for discussion. No clear evidence has been obtained in this respect.

We have described in previous chapters results showing the effect of various substances on gene activation. Some bind more efficiently to single-stranded DNA while others appear to prefer double-stranded DNA [*Frenster*, 1965, 1976; *Neville and Davies*, 1966; *Gamper* et al., 1980] or to intercalate between DNA strands [*Goldberg and Friedman*, 1971; *Le Pecq*, 1978; *Sobell and Jain*, 1972]. This causes physicochemical changes in the DNA which may change the extent of its template activity and, consequently, the replication and/or transcription process. The appearance of single-stranded portions of DNA makes them accessible to DNA or RNA polymerase. This step is necessary for the activity of these enzymes.

On the basis of many data reported in chapters 2, 3, and 5 and our more recent observations [*Beljanski* et al., 1981a; *Le Goff and Beljanski*, 1981, 1982] gene activation might be defined in the following manner:

By binding to DNA or by removing already bound substances from DNA, a given biological or chemical agent may cause the double strand of DNA to open. This allows the transcription of one of the two chains. In the case of strands which have already been locally opened, an agent should bind to only one of them, thus permitting the other strand to be transcribed for a longer period. In contrast, the closing of DNA strands should lead to gene inactivation. This may be performed during cell-to-cell contact and secretion of molecules which bind, for example, to each of the single strands. It may also occur under the influence of some exogenously introduced molecules. Under physiological conditions, gene activation should proceed in coordination with other cell constituents. The opening of the DNA chains for gene expression should be regulated by factors elaborated in the cytoplasm or nucleus or even by exogenous molecules introduced into the organism via nutrients, water, etc., whose effects will be analyzed and discussed here. The excessive activation of a region of DNA beyond a certain limit might lead to an imbalance of various biochemical pathways and this would be dangerous for the cell. In fact, whereas controlled activation leads to increased template activity, activation over this limit, under the effect of a carcinogen, for example, leads to an unphysiological metabolism. Several overactivated genes side by side (as occur during polychemotherapeutic treatment of cancer cells) may lead to complete DNA strand separation and the death of the cell.

Activation of Overlapping Genes

Theoretically, genes should be limited in size. On the basis of molecular hybridization between cloned DNA and corresponding mRNA, the length of a gene may be estimated. This view is further supported by analysis of specific proteins synthesized in vitro using mRNA as template. The possibility of overlapping genes in DNA has, however, been considered in a theoretical model, called 'surimpression des gènes' [*Grassé,* 1973]. This model suggests that a segment of DNA corresponding, in principle, to one gene might be transcribed into two proteins depending on how the segment is read by RNA polymerase. Although this enzyme should start transcription through binding to the 'promoter' region, nothing seems to prevent RNA polymerase to start the transcription on different nucleotides outside of this region. Thus, a given DNA segment may function differently several times during the life of an organism. We have seen that certain stimuli (enzymes, hormones, carcinogens, drugs, RNA, etc.) may determine the limit of a gene by forcing it to produce a specific mRNA. In DNA, overlapping genes begin their activity successively and the limits of each gene are no longer a barrier for recombination of codons. For example, let us suppose that three adjacent genes A, B and C belong to the genotype of an insect (fig. 16): gene A functions during embryogenesis; gene B functions during metamorphosis; gene C functions during sexual maturity. Outside of these periods the A, B and C genes are silent. Under the influence of a gene-activating substance the codons are combined in such a way as to constitute a fourth assembly, i.e., a D gene containing codons e, k, l, m, n, o, x, y of the example which thereafter 'commands' the expression of a character differing from those which are determined by the A, B and C genes. The validity of this hypothetical example has been confirmed by experimental data showing that a segment of DNA belonging to bacteriophage ΦX174 produces a number of proteins superior to that produced by genes arranged in line and carried by this segment [*Barrel* et al., 1976a, b]. These observations imply that the RNA polymerase initiation sites have been displaced under the effect of the gene-activating agent. This view

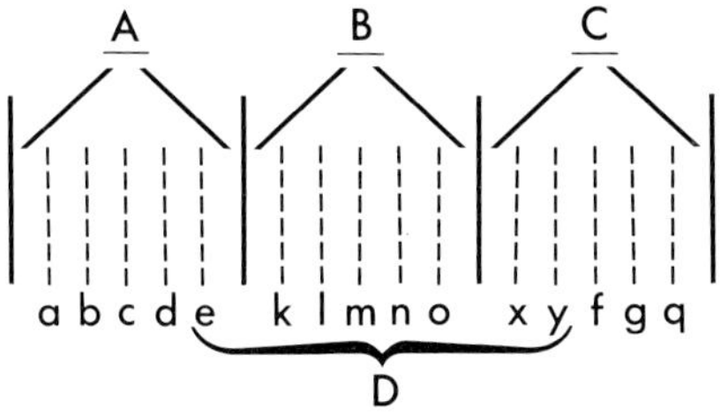

Fig. 16. Superimpression of genes [from *Grassé*, 1973, with permission].

is in agreement with the existence of long transcriptional products from which much smaller 'mosaic' mRNAs emerge. Thus, the dogma of one gene, one enzyme (polypeptide) is no longer valid.

Chemical Agents in the Destabilization of the DNA

Experimental data obtained in vitro and in vivo [*Farber*, 1968; *Frenster*, 1976] have shown that various agents may favor the destabilization of DNA chains in the helical structure, thus creating conditions which permit DNA template activity to occur. Other agents may have the opposite effect, preventing destabilization and subsequent template activity [*Beljanski and Beljanski*, 1982; *Le Goff and Beljanski*, 1982]. Although the detailed operations of these agents are not yet clearly understood, certain experimental data may help to throw light on the question. Various chemical substances such as amines, urea, alcohol, carbonates, etc. bind to DNA by hydrogen bonds and stabilize the denatured secondary structure of DNA [*Levine* et al., 1963]. DNA unwinding by organic solvents (non-intercalating agents) indicates that dehydratation leads to structural perturbation and that water activity in the microenvironment of DNA is related to the stability of the double-stranded DNA [*Lee* et al., 1981]. This structural transformation may result in a change of binding efficiency as well as binding specificity of various agents including carcinogens. Other agents, such as acridines, daunorubicin, adriamycin or psoralen, bind to DNA with non-covalent linkages (they may also intercalate), thus modifying the physicochemical structure of DNA. By using absorption spectra or X-ray diffraction diagrams, evidence may be obtained that under the influence of the above agents DNA molecules become substantially elongated [*Neville and Davies*, 1966]. Daunorubicin, actinomycin D or acridines bind to the external part of DNA (acridine may bind by electrostatic linkage) [*Peackocke and Skerrett,* 1976; *Neville and Davies,* 1966]. They intercalate between DNA strands but may also bind on the surface without preventing the separation of DNA chains [*Marmur and Grossman,* 1961]. In this latter case one of the two DNA strands may be available for transcription. It should be stressed that concentrations of actinomycin D which inhibit the transcription of DNA correspond to doses needed for the increase of the fusion temperature of helicoidal DNA. These concentrations stabilize the secondary structure of DNA with the disappearance of free single strands [*Reich,* 1964]. Under certain conditions, alkylating agents such as mitomycin, BP, or DMBA, form covalent linkages with DNA, thus inducing the modification of the primary and secondary DNA structure. It has been shown that the diolepoxide of BP alkylates DNA (guanine and adenine), resulting in the destabilization of eight base pairs around methylated bases. This alkylation may be diminished by 90% in the presence of Na^+ or Mg^+ [*Gamper* et al., 1980] because these ions protect the phosphates in DNA and favor the winding of DNA chains. This diminishes the extent of the rupture of the hydrogen bonds [*Anderson and Bauer,* 1978]. It should be pointed out that alkylating agents may destabilize in vitro the DNA in the absence of 'activating enzymes', as we recently showed using purified cancer DNAs [*Beljanski* et al., 1981a; *Le Goff and Beljanski,* 1981, 1982]. Thus, without being transformed into diol epoxide, BP causes the in vitro separation of the strands of cancer DNA, followed by an increase in cancer DNA synthesis. It has been observed during tumorigenesis of the skin by DMBA [*Gelboin and Klein,* 1964] that RNA synthesis precedes DNA synthesis. This indicates that in this system, the carcinogen contributes in allowing the transcription, prior to replication. The physicochemical structure of a DNA may be destabilized through methylation of bases in DNA. In fact, methylation of an amino group in position 6 of adenine results in destabilization of DNA [*Engel and von Hippel,* 1978]. The extent is proportional to the number of methylated adenine residues. Methylation of one base in DNA is accompanied by the rupture of several hydrogen bonds around the base. This gives such DNA a better template which may in turn be potentiated by the presence of carcinogens or other compounds (chapt. 5). Although there appears to be a general correlation between gene expression and undermethylation of bases in DNA, there is no clear-cut evidence that changes in methylation control gene expression. Some recent data seem to show that gene sequences, which are in a DNase I-sensitive conformation and are transcribed to RNA are relatively undermethylated at all CpG residues [*Naveh-Many and Cedar,* 1981; *Christman* et al., 1977].

In this connection it should be kept in mind that small oligoribonucleotides bind to DNA and influence the template activity for replication or transcription. They bind to different places on DNA, and their base composition varies according to whether they belong to healthy or cancer cells. This observation is supported by others showing that some RNA fragments which are active primers for DNA replication of healthy cells are not active for cancer cell DNA replication [*Beljanski* et al., 1981a; *Plawecki and Beljanski*, 1981; *Le Goff and Beljanski*, 1979]. However, they may bind to cancer DNA and prevent its replication or transcription.

In vitro Local Opening of DNA Chains

The observations that carcinogens or steroid hormones may induce the appearance of cancer cells and/or the differentiation of particular cell lines [*Lacassagne*, 1936; *Friend* et al., 1971] suggested that, under certain conditions, the secondary structure of DNA chains may be destabilized by such agents. This view was strengthened by theoretical considerations, indicating that the appearance of cancer cells requires the persistent and cumulative action of carcinogens [*Braun*, 1972] or estrogens [*Lacassagne*, 1936] in the systems. In order to achieve DNA destabilization and/or strand separation, hydrogen bonds which hold together DNA complementary chains have to be somehow broken. This can be accomplished at high temperature or alkaline pH, or by the action of various substances, such as enzymes, carcinogens, hormones, RNA fragments, antimitotic drugs, etc. Free purine and pyrimidine bases which interact with DNA destabilize the secondary structure of DNA, and this results in a drop of several degrees in the denaturation temperature [*Ts'o* et al., 1962a, b]. This observation is of particular interest in view of the fact that purines or their analogs induce differentiation of erythroleukemic cells [*Gusella and Housman*, 1976; *Tanaka* et al., 1975]. Once again, differentiation requires release of specific information and this implies local DNA strand separation. It should be pointed out that even cross-linking agents (bifunctional alkylating agents, for example) or intercalating agents will not prevent the strands of the DNA double helix from being separated from each other [*Marmur* et al., 1961].

Destabilization of DNA may be controlled either by a decrease in the denaturation temperature of DNA or by an increase in UV absorbance (hyperchromicity) at 260 nm. The UV absorbance technique is a rather simple one and easily accessible. In order to determine the effect of a compound in DNA strand separation (cancer DNA in particular) at physiological pH (7.3–7.6), five prerequisites are necessary: (1) The DNA to be used must be practically devoid of proteins and RNA, and its integrity must be controlled by ultracentrifugation or by gel electrophoresis. (2) The agents used for strand separation should not cleave the phosphodiester bonds of DNA chains. (3) The effect of an agent on DNA strand separation must be followed in a low-buffered solution at physiological pH. (4) High salt concentrations should be avoided (Na^+, K^+, Mg^{2+}, etc.) since these bind to the phosphates of DNA and neutralize their charge, thus stabilizing the double-helical structure which in turn handicaps the action of agents active in DNA strand separation. (5) Hyperchromicity induced by alkali must be at least between 35 and 45% in order to show the existence of the double-helical structure of DNA.

In our studies on DNA strand separation we used DNAs purified from animal, human and plant tissues, both cancerous and normal, and some bacteria and their mutants [*Beljanski,* 1979; *Beljanski* et al., 1981a, b; *Le Goff and Beljanski,* 1981, 1982]. The main reasons for using these DNAs are as follows: (a) a carcinogen recognizes and activates certain gene(s) in cancer cells or bacterial mutants [*Benedict* et al., 1977]; (b) DNA from many cancer tissues is a much better in vitro template than DNA from healthy tissues, and (c) in the presence of a carcinogen, certain antimitotic drugs or steroids (for cancer steroid target DNA), the in vitro synthesis of cancer DNA increases significantly while that of normal DNA is only slightly enhanced. This last observation suggested that cancer DNAs have a destabilized structure compared to DNA from healthy tissues. The consequence of this might be expressed in the process of replication or transcription. In Chinese hamster embryo cells, transformed by simian virus 40 (SV40), a chemical carcinogen (7,12-DMBA) induces ‘in situ’ preferential amplification of the heterogeneous collection of SV40 DNA inserts [*Lavi,* 1981]. Some of the excised DNA segments containing a site for the origin of replication may be further amplified as independently replicating genomes. It should be stressed that in the above cells carcinogen action does not lead to the rescue of an infectious section of a virus or a complete viral DNA.

To test the fragility of hydrogen bonds in DNA, we performed the in vitro assay, which means following the UV absorption of a given DNA at 260 nm in the absence or presence of a carcinogen. Thus, when 9,10-DMBA (10 μg/ml) is added to a solution containing DNA from healthy tissues, the increase in UV absorbance (hyperchromicity) is low but detectable (about 3%). When DNA from a corresponding cancer tissue is used under the same experimental conditions, the UV absorbance increase is between 15 and 30%

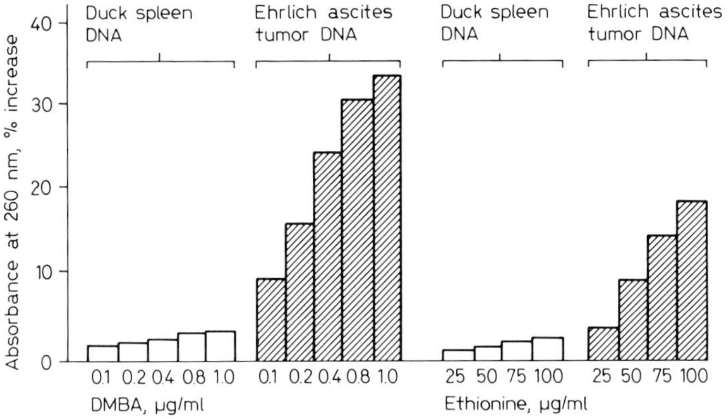

Fig. 17. Ehrlich ascites tumor DNA and spleen DNA strand separation. Effect of DMBA or ethionine. UV absorbance of DNA was measured in the absence or presence of each compound [from *Beljanski* et al., 1981a].

(fig. 17). This latter increase can be repeatedly obtained with numerous cancer DNAs. We have reported that 1-(2-chloroethyl)-*l*-nitrosourea (CCNU), ethionine, daunorubicin, 5-fluorouracil (5-FU), actinomycin D, mitomycin, steroids and other substances induce local strand separation in cancer DNA [*Le Goff and Beljanski*, 1981; *Beljanski* et al., 1981a] from mammalian and plant tissues, each being effective in a given concentration. The effect of high doses may cause the DNA to collapse.

The effect of steroid hormones on DNA from hormone target tissues is of particular interest from the theoretical and practical points of view, i.e., carcinogenesis and anticancer therapy. We have recently shown that both testosterone and estradiol, separately incubated in vitro with DNA from normal breast tissue or from breast cancer, have a low hyperchromic effect (1–3%) on breast DNA while in breast cancer DNA the values are 5–8% with estradiol and 18% with testosterone. There are optimal concentrations for DNA strand separation. DNA isolated from tissues which are not steroid targets does not respond to steroids as far as strand separation is concerned [*Beljanski* et al., 1981a]. However, surprisingly, DNA from human neurocarcinoma appears to be an exception. In fact, this DNA reacts to hormones in the same way as DNA from hormone target cancer tissues. It should be noted that steroid hormone receptors have been found in tissues considered non-target tissues, such as brain or liver (chapt. 3, 5).

Figure 18 shows the percentage increase of UV absorbance in neurocarcinoma DNA when incubated in vitro with estradiol. This increase can be

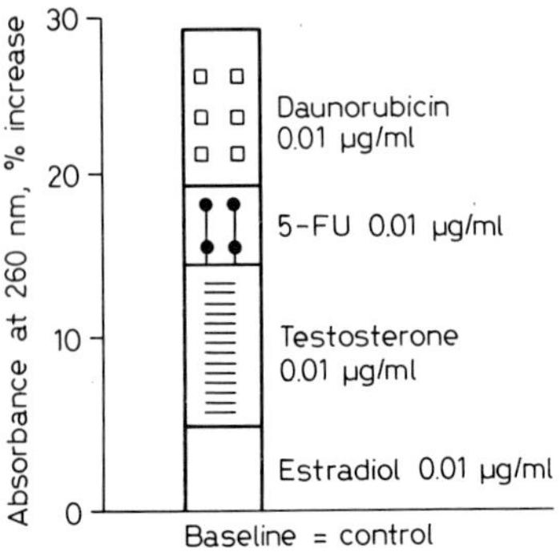

Fig. 18. Human neurocarcinoma DNA strand separation. Effects of estradiol, testosterone, 5-FU and daunorubicin [from *Beljanski* et al., 1981a].

substantially augmented again by successive additions of testosterone to the incubation mixture followed by 5-FU and finally daunorubicin. Each of these agents induces a certain degree of local DNA chain separation, and their added effects lead to hyperchromicity of about 30% which is considerable since the corresponding figure in the presence of KOH is about 42–44%. The cumulative and limited effect of each of these agents suggests the presence of different binding sites on DNA, the nature of which is not yet known. The fact that DNA from cancer tissues responds efficiently to the solicitations of various agents while DNA from healthy tissues responds poorly, shows the existence of accessible binding sites on cancer DNA. We may ask whether the fact that locally separated DNA segments are present in greater number in cancer DNA than in DNA from normal cells has something to do with Z-DNA structure. This left-handed DNA has in its backbone an irregular or zigzag configuration in which few CpG residues are found on the outside of the molecule [*Wang* et al., 1979]. Low-level alkylation of DNA with carcinogens may be related to the formation of Z-DNA structure in vivo. However, one cannot exclude the possibility that other molecules such as peptides, proteins or antimitotics may convert right-handed DNA (B-DNA) to Z-DNA. It should be noted that a segment of the histidine D gene of *Salmonella* has a mutational 'hotspot' which occurs at a segment of eight G-C residues [*Isono and Yourno,* 1974]. Our investigations have shown that DNA from mammalian cancer cells and DNA from His$^-$ *Salmonella* mutant are much more susceptible to carcinogens or antimitotics than DNA from healthy cells or from wild strains of *Salmonella typhimurium* [*Beljanski* et al., 1981a, 1982a].

It is conceivable that chemotherapeutic drugs play a role in separating the strands of cancer DNA above a certain threshold. Once this has been done, the pathways for synthesis are disconnected and the cell is expected to die. However, these drugs also act on normal cells causing severe damage. Their persistent and progressive action on DNA strand separation may, therefore, overpass a threshold [*Wolsky,* 1978] and cause cancer cells to appear. Chemical and biological stimuli may determine the zones of open DNA chains which may facilitate the binding or the displacement of different physiological molecules in order to favor DNA transcription.

DMSO, Croton Oil and Phorbol Oil Derivatives and DNA Strand Separation

The decision to study the effects of the above agents on DNA strand separation was dictated by numerous observations showing their capacity – when used at appropriate concentrations under the appropriate conditions – to induce release of specific information in cancer cells, resulting in RNA synthesis or cell differentiation (chapt. 5). Thus, DMSO might allow DNA replication in some cells, cultured in vitro, without inducing cell differentiation [*Collins* et al., 1978]. However, this agent induces cell differentiation in erythroleukemic cells [*Friend* et al., 1971; *Blau and Epstein,* 1979]. It should be recalled that DMSO, frequently used as a solvent, denatures in vitro DNA or RNA [*Herskowits,* 1962; *Iglewski and Franklin,* 1967; *Katz and Penman,* 1966] and also dissociates the 'DNA-RNA polymerase complex' (chapt. 2). Croton oil or phorbol oil and its derivatives play an important role in cell differentiation and in the promotion of carcinogenesis in mammalians [*Gelboin and Klein,* 1964; *Berenblum,* 1941; *Kopelovich* et al., 1979]. No explanation of the action of these agents was put forward in these studies. DMSO (fig. 19), croton oil and phorbol oil strongly enhance both cancer and normal DNA synthesis in vitro, in roughly similar proportions. These data indicate that in vitro they do not distinguish cancer DNA from healthy tissue DNA as do carcinogens or steroids [*Beljanski,* 1979; *Le Goff and Beljanski,* 1981]. At room temperature, low concentrations of DMSO cause a progressive increase in UV absorption of DNA both from healthy and from cancer cells. In the presence of croton oil, neurocarcinoma DNA and monkey brain DNA show about 40% increase of UV absorbance. This increase is very high when compared to the increase observed when these DNAs were incubated with alkali (in which case hyperchromicity was of the order of 42–44%).

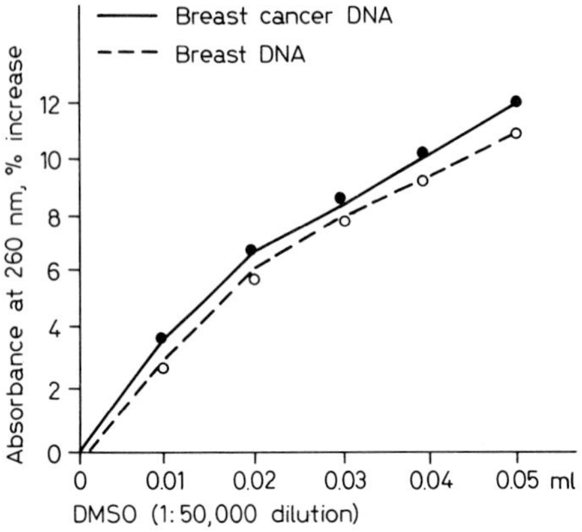

Fig. 19. Human breast cancer and normal DNA strand separation. Effect of DMSO [from *Beljanski* et al., 1981a].

At pH 7.60, croton oil exhibited little effect on either cancer or normal DNA strand separation, while at pH 7.65 or 7.70 [*Beljanski* et al., 1981a] its effect was very substantial on both types of DNAs. We should stress that croton oil induces separation of strands in healthy monkey spleen DNA, which might be amplified by the addition of 9,10-DMBA [*Beljanski,* unpublished results]. On the basis of these experiments, it would appear reasonable to interpret a great many experimentally established data concerning the effect of DMSO or phorbol oil, in cell differentiation or in the liberation of specific information, on local DNA strand separation in cells cultured in vitro. It has recently been reported that chronic lymphatic leukemia cells, which are weak producers of interferon, synthesize 3–4 times more interferon after treatment with TPA isolated from phorbol oil and thus equal the amounts produced by normal cells [*Adolf* et al., 1981]. No interpretation of TPA action was suggested in this report. Although DMSO was used as solvent for TPA, its possible effects were not controlled. It should be mentioned that PMA or TPA induces cancer DNA strand separation locally [*Beljanski and Le Goff,* in press]. This might explain the observations just mentioned concerning interferon production.

PMA or TPA has effects on DNAs from some healthy cells. This indicates that the phorbol oil from which they have been isolated contains agents which are active in healthy cell DNA strand separation. The extent to which cancer DNA chains are opened (local or complete opening) is dose-dependent in the presence of these compounds as well as in the presence of DMSO or actinomycin. It should be recalled that specific concentrations of DMSO or actinomycin induce differentiation of leukemic cells into macrophages or granulocytes [*Lotem and Sachs,* 1979]. Their action may be seen on different segments of the same DNA molecule. The mechanism which all of the above agents share would appear to be the capacity to open DNA strands. Phorbol esters induce and stimulate the proliferation or differentiation of not only cancer cells but also of normal cultured cells. Thus, both TPA and phorbol 12,13-didecanoate stimulate myeloid colony growth, which depends on the incubation medium. Only a temporary exposure (45 min) of mouse bone marrow cells to phorbol esters is necessary to produce the growth-stimulating effect. In contrast, similar concentrations of these compounds inhibit the formation of colonies of early erythroid progenitor cells [*Sieber* et al., 1981]. No explanation for the exact mechanism by which phorbol esters affect myelopoiesis was provided. It should be stressed that phorbol esters, among them TPA, induce local DNA strand separation, thus allowing accelerated DNA replication and, consequently, cell division. The fact that the presence of fetal calf serum in the culture medium is necessary suggests that RNase present in the serum may provide RNA primers which in cooperation with phorbol esters accelerate DNA replication and transcription in certain cells. For other cells they might have an inhibitory effect.

Opening and Closing of DNA Chains and Initiation of DNA Replication

According to experimental data reported in various publications, carcinogens appear to increase the number of sites implicated in the replication of DNA [*Walters* et al., 1976; *Lown* et al., 1978; *Kato and Strauss,* 1974]. These data imply that carcinogens do not bind directly to the initiation sites but rather to areas close by. It has been shown that low concentrations of hydroxyurea enhance the relative proportion of initiation of replication to elongation in the DNA of HeLa cells. It invariably causes at least a doubling of newly synthesized DNA with unusual base-pairings [*Nilsen and Baglioni,* 1979]. Specific leader sequences preceding the initiation sites at the level of replicon origin in mammalian cells are important as shown for SV40 DNA which has a unique nucleotide sequence [*Dhar* et al., 1977]. Carcinogens appear to bind to this sequence preferentially. Since the DNA within the eukaryotic chromosome is heterogeneous in both base composition and base sequence arrangement, it makes interaction sites along the chromosomes

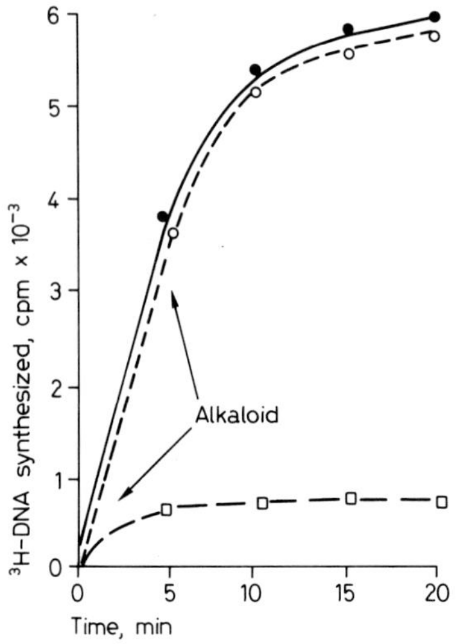

Fig. 20. Inhibition of the initiation of breast cancer DNA synthesis in vitro by alkaloid (serpentine) [from *Beljanski and Beljanski,* 1982].

accessible to specific agents [*Johnston* et al., 1978]. To demonstrate the validity of the above observations we studied the in vitro competition between particular alkaloids which bind specifically to cancer DNA replication sites and carcinogens which increase the number of initiation sites in DNA from cancer cells.

Figure 20 shows that alkaloids of the β-carboline class, such as alstonine or serpentine, sempervirine or flavopereirine, block the initiation of cancer DNA synthesis but have no effect on DNA chain elongation. Furthermore, the carcinogens which liberate the initiation sites on DNA presumably allow the binding of alkaloids at any moment during their stimulating action of cancer DNA synthesis. If the initiation sites are liberated, the alkaloids should rapidly stop DNA synthesis. This has been shown to be the case [*Beljanski and Beljanski,* 1982]. Figure 21 shows that the addition of alkaloids to the DNA-synthesizing medium at the moment of increased DNA synthesis in the presence of a carcinogen results in prevention of further DNA synthesis. These results strongly suggest that in vivo alkaloids of the β-carboline class may act as preventive agents and suppress the stimulatory effect of carcinogens, provided that the appropriate concentrations are used. There appears to be a competition between the alkaloids and carcinogens tested (ethionine, DMBA, reserpin). Besides carcinogens, we studied the effect of

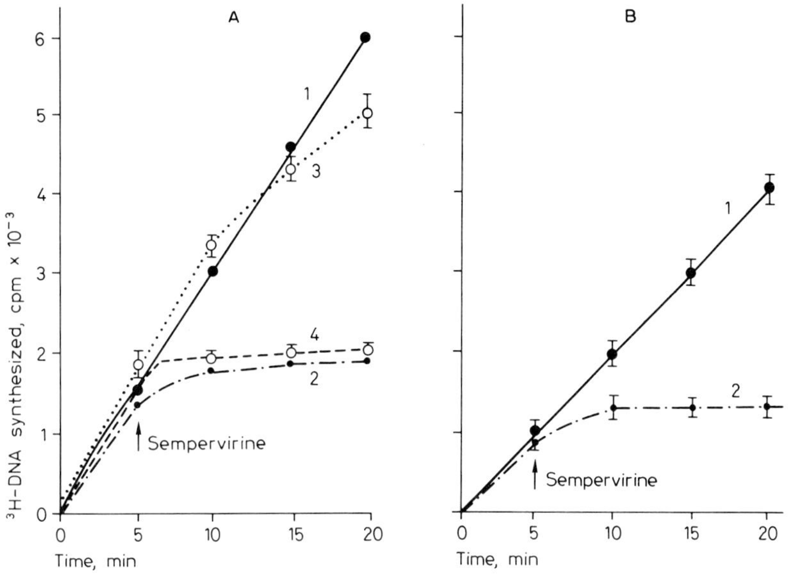

Fig. 21. Effect of sempervirine on cancer DNA synthesis in vitro in the presence of testosterone, *dl*-ethionine or 7, 12-DMBA. For incubation conditions, see table I, footnote. *A* Lung DNA (0.25 μg) was used as template under two conditions: 1 = Complete medium + ethionine (60 μg) at time 0 (control); 2 = complete medium + ethionine (60 μg) at time 0 and sempervirine (25 μg) at the 5th min of incubation; 3 and 4 = breast cancer DNA (0.25 μg) was used as template; 3 = complete medium + testosterone (40 μg) at time 0; 4 = complete medium + testosterone (40 μg) at time 0 and sempervirine (25 μg) at the 5th min of incubation ($p < 0.001$, significance value in comparison to control, for all points except for 10 min: $p < 0.01$. *B* Breast cancer DNA (0.25 μg) was used as template. 1 = Complete medium + 7,12-DMBA (60 μg) at time 0 (control); 2 = complete medium + 7,12-DMBA (60 μg) at time 0 and sempervirine (25 μg) at the 5th min of incubation ($p < 0.001$, significance value in comparison to control) [from *Beljanski and Beljanski*, 1982].

steroid hormones and alkaloids on DNA isolated from human breast tissue and corresponding cancer tissue, respectively. The possible competition between these substances during in vitro DNA synthesis was also studied. Alkaloids inhibit the initiation of breast cancer DNA synthesis but not chain elongation (fig. 20). They have no effect on DNA synthesis from healthy breast tissue. Steroids, which are non-mutagenic hormones, act as regulators in the activity of hormone target cells [*Cox*, 1980] and strongly stimulate breast cancer DNA synthesis in vitro (fig. 21) [*Beljanski* et al., 1981a]. Alkaloids prevent this enhancement in the presence of the steroid hormones. However, when hormones are used at high concentrations they overcome the

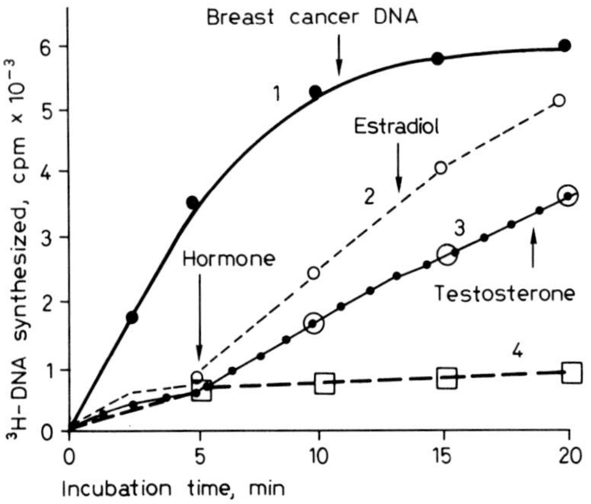

Fig. 22. Steroids may overcome breast cancer DNA synthesis inhibited by serpentine. For incubation conditions, see table I, footnote. Breast cancer DNA (0.5 μg) was used as template. 1 = Complete medium; 4 = complete medium + serpentine (80 μg) at time 0; 2 = complete medium + serpentine (80 μg) at time 0 and estradiol (50 μg) at the 5th min of incubation; 3 = complete medium + serpentine (80 μg) at time 0 + testosterone (50 μg) at the 5th min of incubation [from *Beljanski and Beljanski,* 1982].

inhibiting effect of the alkaloid (fig. 22), suggesting that steroids might liberate the initiation sites on breast tissue DNA, i.e., sites to which alkaloids bind. These events were not observed when breast cancer DNA was replaced by several DNAs from healthy or cancer tissues which were not steroid targets. Although steroid hormones, as well as carcinogens, liberate the initiation sites on breast cancer DNA, their binding sites on DNA are not the same as those to which carcinogens bind. Our preliminary data show that there is no competition between carcinogens and steroids during breast cancer DNA synthesis in vitro. The nature of the binding sites is not known.

Alkaloids of the β-carboline class which bind to DNA initiation sites on cancer DNA might prevent RNA primers from activating DNA replication. However, their inhibitory effect could be reversed by an excess of P_1, P_2 primers [*Beljanski* et al., unpublished results]. BLR primers which act in vitro and in vivo in the replication of normal cell DNA are inactive as primers for cancer cell DNA and do not activate the multiplication of these latter in vivo [*Beljanski* et al., 1981b; *Plawecki and Beljanski,* 1981].

The alkaloids mentioned bind in vitro readily to human breast cancer DNA but poorly to DNA from healthy cells (fig. 23). This differential binding may be due to the existence of local areas where two strands are unpaired

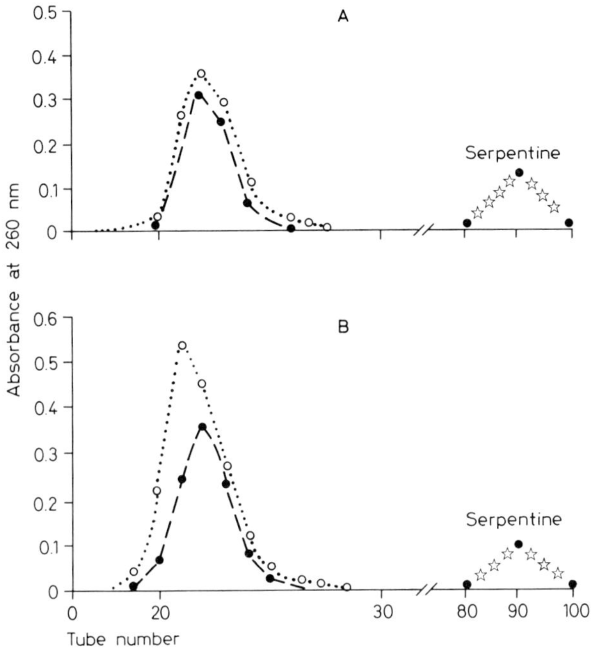

Fig. 23. Cancer DNA-serpentine complex in in vitro formation. *A* ● = Breast DNA; ○ = breast DNA + serpentine. *B* ● = Breast cancer DNA; ○ = breast cancer DNA + serpentine. [From *Beljanski and Beljanski,* 1982.]

on cancer DNA. The unpaired strands probably carry initiation sites for DNA replication, to which all of the above alkaloids specifically bind. Differential binding of alkaloids to cancer DNA must be interpreted by taking into account the following observations:

(1) Cancer DNA is a better template for the in vitro synthesis of DNA than DNA from healthy cells. DNA polymerase requires single strands as template.

(2) Carcinogens stimulate cancer DNA synthesis in vitro strongly and that of DNA from healthy cells poorly.

(3) The UV absorbance increase of cancer DNA (at 260 nm) in the presence of carcinogens shows that cancer DNA is destabilized compared to DNA from healthy tissues.

(4) Psoralen, which intercalates between the strands of cancer DNA, may be displaced by the same alkaloids [*Beljanski*, unpublished results], which suggests that alkaloids bind to DNA by intercalating between DNA chains.

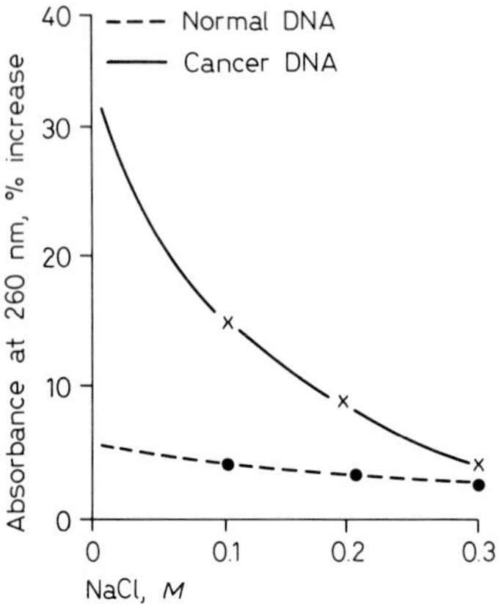

Fig. 24. Cancer and normal DNA strand separation in the presence of 9,10-DMBA. Effect of ionic strength. Ehrlich ascites tumor DNA (20 μg) and normal mouse or monkey spleen DNA (20 μg/ml) were each dissolved in 1 ml of 10^{-2} *M* HCl-Tris buffer (pH 7.65) containing various concentrations of NaCl as indicated. For each NaCl concentration 10 μg of DMBA were used. The absorbance was measured at 260 nm. The DNA samples were read against a blank cuvette containing NaCl at given concentrations and 10 μg of DMBA/ml [from *Beljanski* et al., 1981a].

(5) NaCl, $MgCl_2$, KCl, and other salts may overcome the stimulation of cancer DNA synthesis by carcinogen (fig. 24). Salts by binding to DNA phosphates contribute to stabilizing hydrogen bonds and prevent DNA chain openings. This results in the absence of fixation of alkaloids to cancer DNA.

Carcinogenic Agents and the Multiplication of Mammalian and Plant Cancer Cells

The aim of this line of research was to correlate the various observations made in vitro on carcinogenic agents stimulating cancer DNA synthesis and cancer cell multiplication in mouse tissue and then compare these phenomena to carcinogenesis in plant tissue. Increased DNA synthesis and strand separation in vitro should be equivalent to increased cancer cell multiplication in vivo in the presence of these carcinogens. We have tested the effect of DMBA and CCNU on tumor development in mice inoculated with Ehrlich ascites tumor cells. In order to activate DNA synthesis to the full extent it is necessary to achieve complete DNA strand separation for replication of the whole genome and thus induce cell division. Transcription of DNA and

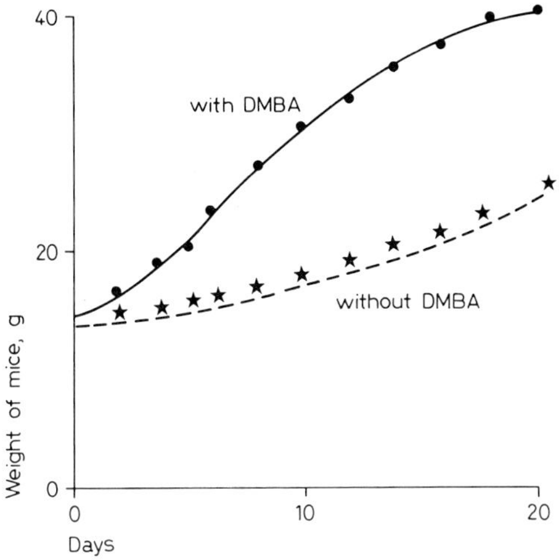

Fig. 25. Stimulation of Ehrlich ascites cells in mice by DMBA [from *Beljanski* et al., 1981a].

translation should be also accelerated. To achieve this, one group of mice received low doses of 9,10-DMBA in the tumor inoculation site for 5 consecutive days (intramuscular inoculation was performed to put the agent in direct contact with the tumor cells). As shown in figure 25, the average weight of DMBA-treated mice rapidly increased compared to controls. On the 20th day the mean tumor weight of control mice was 4.5 ± 0.03 g (mean ± SD) and that of DMBA-treated mice was 10.8 ± 0.5 g. This experiment was duplicated with CCNU, but in this case the mice were inoculated intraperitoneally. The mean weight increase of CCNU-treated mice (100 μg of CCNU per mouse, administered intraperitoneally every 2nd day over a period of 15 days) was 10.5 g and that of control mice was 6 g. These experiments show a reasonable degree of correlation between the stimulating effects of DMBA and CCNU on in vitro cancer DNA-accelerated synthesis, cancer DNA strand separation and in vivo multiplication of cancer cells [*Beljanski* et al., 1981a].

A relatively close relationship exists between carcinogenesis in the animal and plant kingdom. Certain chemical carcinogens do induce both mammalian cancer [*Arcos and Argus*, 1974] and plant tumors [*Bendar and Linsmaier-Bendar*, 1971; *Garrigues* et al., 1971]. Low concentrations of carcinogens stimulate multiplication of mammalian as well as plant cancer cells

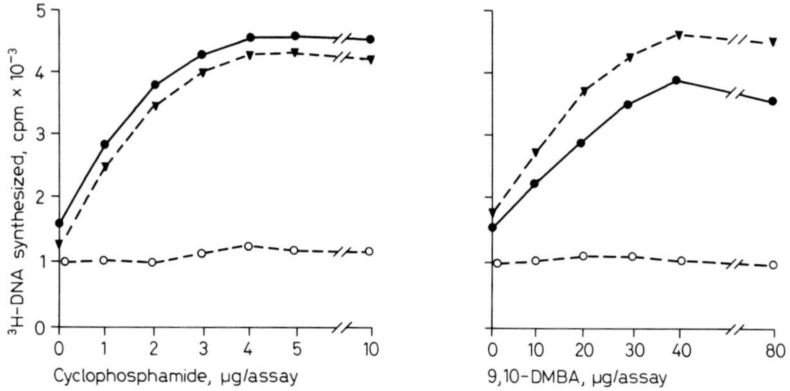

Fig. 26. Effects of cyclophosphamide and 9, 10-DMBA on cancer and normal DNA synthesis in vitro ● = Crown-gall DNA; ▼ = *A. tumefaciens* DNA; ○ = healthy pea cell DNA [from *Le Goff and Beljanski,* 1981].

[*Bear and Schneider,* 1979]. We have recently demonstrated [*Le Goff and Beljanski,* 1979, 1982] that a correlation exists between crown-gall DNA synthesis in vitro, DNA strand separation and multiplication of cancer cells of pea seedlings (induced by *Agrobacterium tumefaciens,* oncogenic strain B_6). Cyclophosphamide, daunorubicin and DMBA either substantially accelerate or inhibit the multiplication of cancer cells. When introduced into wounded tissues of germinating and decapitated *Pisum sativum* as early as 24 or 48 h following infection with *A. tumefaciens* B_6 (a 24-hour lapse of time is needed for conversion of healthy plant cells into tumor cells), low concentrations of cyclophosphamide, daunorubicin or DMBA stimulate cancer cell multiplication. 4 days after inoculation of the bacteria, stimulation of the neoplastic cell growth by the three agents can still be observed. The compounds at the same low concentrations have no effect when introduced at the same time as the bacteria, i.e., at the time of cancer induction. The lack of action at time 0 may be the consequence of the dilution or elimination of these substances unretained by healthy cells during the 1st hour following the injury. These results correlate the stimulation of crown-gall DNA synthesis in vitro and the absence of stimulation on DNA synthesis from healthy cells (fig. 26). At high concentrations, each of these three agents shows a strong inhibiting action. It must be emphasized that in the absence of bacteria, none of these compounds has a detectable carcinogenic effect on plants under the experimental conditions used. This indicates that tumor induction is essen-

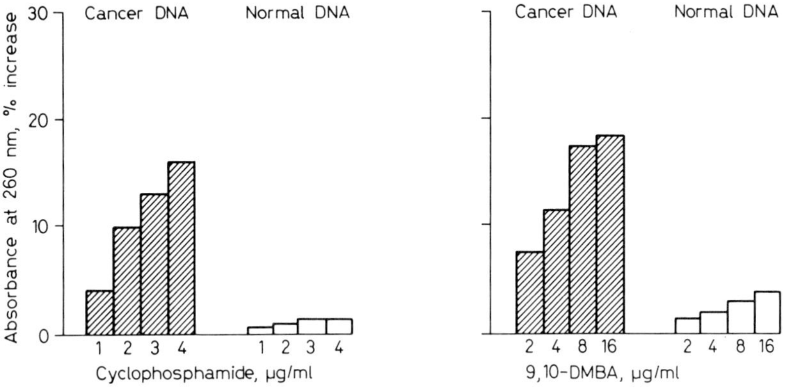

Fig. 27. Effects of drugs on crown-gall tumor and healthy pea cell DNA strand separation. UV absorbance (260 nm) of DNA was measured in the absence or presence of each compound tested at the indicated concentrations [from *Le Goff and Beljanski,* 1981].

tially due to the effect of *A. tumefaciens* and not to carcinogens. However, the carcinogens used at the given concentrations stimulate the multiplication of crown-gall tumor cells in plants.

The substantial and selective stimulation of in vitro crown-gall DNA synthesis and the significant increase of tumor cell multiplication in whole plants under the influence of each of these compounds suggest that the cancer DNA was 'relaxed' compared to normal DNA and was ready to undergo further destabilization in the presence of these substances. In fact, data illustrated by figure 27 show that in the presence of each of these agents, UV absorbance considerably increased with crown-gall DNA whereas the increase was very slight for control cell DNA. There are optimal concentrations for DNA strand separation which vary according to the drug used (fig. 27). They also locally separate the strands of DNA from *A. tumefaciens* B_6 but have no effect on the strands of DNA from *E. coli* wild strain or nononcogenic B_6Tr1 strain [*Beljanski* et al., 1972a]. The DNAs of these latter are in a stabilized condition as may be judged by their low template activity for DNA synthesis and the poor response to carcinogens used to stimulate DNA synthesis or DNA strand separation. On the basis of the observations described above, the following general conclusions can be drawn:

(1) Cancer DNA from mammals and plants have a secondary structure which is destabilized compared to DNA from healthy tissues.

(2) Carcinogens used at low concentrations further destabilize the secondary structure of cancer DNAs in vitro.

(3) Carcinogens strongly stimulate the in vitro synthesis of cancer DNA but only slightly that of DNA from healthy cells.

(4) Low doses of carcinogens stimulate both mammalian and plant cancer cell multiplication in situ.

Possible Causes of the Destabilized Condition of DNA in Cancer Cells

The fact that cancer DNAs from mammals and plants are destabilized compared to DNAs from healthy tissues prompted us to visualize the effect of carcinogens and related agents on DNAs which is different from causing mutations. This approach was supported by the following facts:

(a) Non-mutagenic steroids induce the appearance of cancer cells in mammalian steroid-target tissues (chapt. 3).

(b) Sequence analysis of cancer and normal DNA [*Le Pecq,* 1978] and corresponding mRNA [*Myozis* et al., 1980] shows no difference between cancer and normal DNA or corresponding mRNAs.

(c) Increased cancer cell multiplication in animals is correlated with strand separation and enhanced synthesis of cancer DNA in vitro in the presence of carcinogens. This was observed also with steroid hormones for hormone target tissue DNA.

(d) The appearance of cancer cells in mammals is not sudden but results from the persistent and cumulative effects of carcinogens on healthy cells [*Lacassagne,* 1936; *Braun,* 1972].

(e) Some cancer cells may become differentiated or transformed into normal cells by DMSO, carcinogens or tumor promoters.

Normal plant cells may become transformed into 'anergic tissue', i.e., tissue exhibiting several characteristics of crown-gall tumors by persistent use of hormone for in vitro cultivation. Such tissues possess a destabilized DNA [*Le Goff and Beljanski,* 1981, 1982]. The maintenance of the destabilized form is probably due to the binding or removal of some peptides, amino acids, small-size RNA, or similar constituents, whose nature and amount differ from those present in normal DNA. It has been reported that particular peptides found associated with DNA in the nucleus of cow thymus cells bind in vitro to DNA and make the double-helical structure of it extremely stable. Furthermore, the amount of peptides bound to the mRNA of cancer cells is reduced by about 40% in comparison to that found attached to the mRNA of normal cells. Their reduced level in cancer cells may be due to proteases whose activity increases under the influence of promoters of carcinogenesis [*Heinrich* et al., 1978; *Hillar and Przyjenski,* 1973]. The production of hemoglobin in erythroleukemic cells, stimulated by DMSO, is substantially reduced in the presence of these peptides. In addition, arginine and lysine-rich peptides and free amino acids also interact with DNA to form hydrogen bonds [*Lancelot,* 1977a, b; *Lancelot and Helene,* 1977]. Bases of nucleic acids and amino acids have different polar groups capable of forming at least one hydrogen bond with each of the four bases. Glutamic and aspartic acid form a specific complex with guanine only, while arginine may form a complex both with cytosine and guanine. Polypeptides containing 30–100% of lysine significantly reduce the number of binding sites for actinomycin D on healthy cell DNA. The effect of polypeptides depends on the presence and distribution of lysine residues in the chain. Among hydrophilic amino acids present in the peptides only leucine prevents the fixation of actinomycin D to DNA [*Votanova* et al., 1978]. Our preliminary experiments [*Beljanski* et al., unpublished results] indicate that DNA, purified from cancer cells, did not contain the same peptides as DNA from healthy cells, and the peptide concentrations also differ. We have shown that actinomycin D separates the chains in cancer DNA but not, or not readily, in 'healthy' DNA. This difference may well be due to peptides bound to DNA. Further study might support this interpretation, but it should be kept in mind that carcinogens or promoters may lead to destabilization of the helical structure of cancer DNA by increasing protease activity. For example, α-thrombin, a proteolytic enzyme, when incubated with quiescent cultures of chick embryo fibroblasts, stimulates DNA synthesis and cell replication in a culture medium deprived of serum [*Perdue* et al., 1981].

The presence of a particular class of amino acids, guanidines, in plant tumor cells induced by *Agrobacterium tumefaciens* and their role in the process is a matter of discussion [*Braun,* 1972]. An interesting observation, reported 10 years ago, has shown that octopine stimulates the multiplication of plant crown-gall tumors while arginine, from which it is derived, is inactive [*Lippincott and Lippincott,* 1970]. It should be stressed that there are plant tumors which contain nopaline, octopine or lyposine (or combinations of two of these three guanidines). Healthy plant cells appear not to contain (or to contain only traces) of these amino acids [*Braun,* 1972]. Thus, crown-gall tumor cell multiplication in the presence of a guanidine offers an interesting possibility for a study of the relationship between the effect of guanidines on plant cancer DNA synthesis and DNA strand separation in vitro and multiplication of tumor cells in vivo. Tumor cells induced by *A. tumefaciens* synthesize one or two guanidines. The bacterium itself in vitro metabolizes these guanidines.

Our studies show that octopine (which originates biochemically from arginine) induces in vitro DNA strand separation in that strain of *A. tumefaciens* which metabolizes only octopine. It has the same effect on DNA isolated from crown-gall tumors induced by this type of bacterium. It has no effect on DNA from the strain of *A. tumefaciens* which metabolizes nopaline or lysine. *l*-Arginine or *l*-lysine cannot replace the octopine or lysine effect on crown-gall DNA from corresponding strains. Guanidines have practically no effect on DNA from healthy plant tissues.

Guanidines strongly stimulate cancer DNA synthesis in vitro, as illustrated in figure 28. They have a slight stimulating effect on DNA from healthy tissues. Octopine strongly stimulates *A. tumefaciens*-induced crown-gall cells in vivo [*Lippincott and Lippincott*, 1970]. Here again, there is a correlation between crown-gall DNA synthesis in vitro, DNA strand separation and multiplication of plant cancer cells. It should be stressed that neither octopine nor nopaline or lysopine have an effect on DNA from mammalian cells, healthy or cancerous, nor on DNA from non-oncogenic bacteria. During induction of plant carcinogenesis specific genes for the production of guanidines are activated and expressed. This implies DNA strand separation and indicates that the product (guanidine) of the newly expressed genes may contribute in maintaining the destabilized helical structure of cancer DNA. These particular markers make plant cancer cells a model for understanding the basic underlying mechanism of tumor induction.

Alkylated mRNAs and Production of Typical Proteins in vitro

We have described conditions in which the template activity of cancer DNA in vitro can be increased by carcinogens. The carcinogens in this case do not act as alkylating agents. Under these conditions, physicochemical changes in the secondary structure of DNA (local chain separation) appear to be of very great importance. This view is supported by observations that there is no detectable difference in nucleotide sequences between DNA from cancer tissues and DNA from corresponding healthy tissues. Also, mRNAs generated by cancer cell DNA appear to have the same nucleotide sequence as mRNAs from healthy cells. These data suggest that if certain basic changes in cancer DNA and in the resulting mRNAs would be due to mutations, they

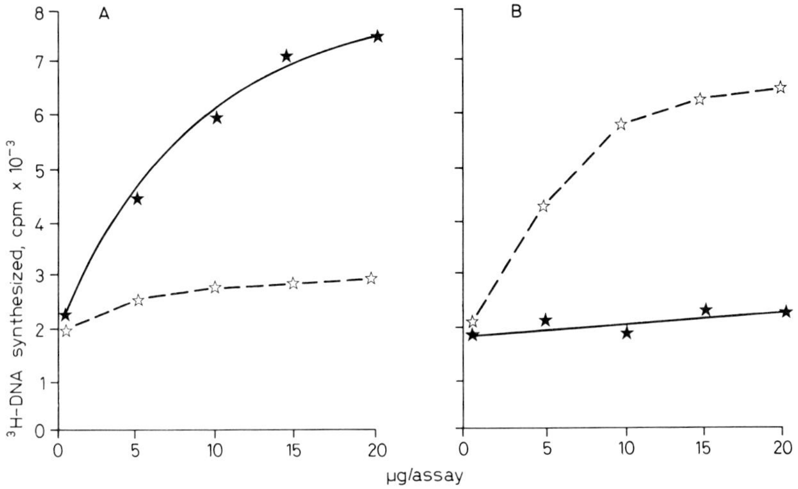

Fig. 28. DNA synthesis in vitro in the presence of either octopine or nopaline. *A* —— = *A. tumefaciens* B_6 DNA + octopine; - - - = *A. tumefaciens* B_6 DNA + nopaline. *B* - - - = *A. tumefaciens* BNV 7 DNA + nopaline; —— = *A. tumefaciens* BNV 7 DNA + octopine [*Le Goff and Beljanski*, unpublished results].

would not be easily detectable by conventional techniques. Experiments were performed recently with the aim of showing that the introduction of alkyl groups into specific mRNA can affect the translating function of mRNA in vitro. We have chosen a few typical cases to illustrate the effect of some alkylating agents on the mRNA function. Diol epoxide of BP, a potent environmental carcinogen and mutagen, was used to modify the activity of globin mRNA in the translation process. With quite low levels of adduct formation (0.4 and 2.4 per molecule of mRNA), globin production was diminished by 50–90%. However, the main substance produced was typical globin [*Grunberger* et al., 1980]. These studies also showed that modification of poly (U_3G) with diol epoxide of BP has no effect on its template activity. Several interpretations were put forward to explain the loss of efficiency observed in carcinogen-treated globin mRNA. The most attractive explanation attributes conformational changes in carcinogen-treated mRNA due to the presence of bulky adducts somewhere on the RNA molecule. It remains possible, however, that a region of 50 nucleotides between the 'cap' region ('cap' = formation and structure of a 7-methylguanosine group in 5′-5′-triphosphate linkage in the 5′-terminal position of many pre-mRNA and mRNA molecules) and the initiator AUG codon in globin mRNA is particularly susceptible to carcinogens since this region is rich in purines [*Efstratiadis* et al., 1977] which are attacked preferentially by diol epoxide. Disruption of the ordered secondary structure of the globin mRNA could make it less recognizable to ribosomes and initiation factors in translation [*Grunberger* et al., 1980]. This interpretation is in agreement with observations showing that considerably higher numbers of methyl or ethyl groups result in lower activity in plant viral mRNA as compared to the bulky hydrocarbons in globin mRNA [*Fraenkel-Conrat and Singer,* 1980]. Introduction of 4–28 alkyl groups per molecule of plant viral mRNA diminishes but does not prevent translation of such RNAs in the wheat germ system. The fact that random alkylation (mostly of either the guanosine N-7 by dimethylsulfate or the phosphate by ethylnitrosourea) does not prevent full-length protein formation shows that alkylations allow initiation and synthesis of specific proteins. The alkyl groups appear not to be located in nonfunctional parts of these mRNAs. The silent gene of BMV-3 seems less silent as a consequence of alkylation. These results are similar to those reported with bisulfite-treated polycistronic MS 2 phage messenger RNA [*Brawerman* et al., 1975]. Treatment of phage MS 2 RNA with BP-diol epoxide impairs the template capacity of this RNA. When MS 2 RNA is alkylated (molar ratio of 10), RNA synthetase formation considerably decelerates, while formation of maturation and coat proteins is only slightly affected. Surprisingly enough, the loss of viral RNA infectivity follows a single-hit inactivation mechanism, whereas several alkylation events in the viral RNA synthetase cistron are required to block translation of this gene product [*Sagher* et al., 1979]. The reduction of the in vitro translation of alkylated plant viral mRNAs might be the consequence of conformational changes due to the loss or gain of negative or positive sites in alkylated mRNA [*Fraenkel-Conrat and Singer,* 1980]. Modified mRNA is less efficient in forming the initiation complex with ribosomes than the unmodified mRNA, but the products are typical viral proteins.

Various Agents Connected with the Activation of Genes

Accelerated DNA strand separation may lead to accelerated cell division. When cell division is accelerated by an agent, transcription of several genes must be accomplished if the cell is to survive and divide. Replication and transcription have to be coordinated. Gene activators may selectively induce transcription of specific genes. Thus, using a single leukemic cell, it was shown that actinomycin D, when used at a very low concentration (2 ng/ml), causes differentiation of the leukemic cell into granulocytes, while TPA or DMSO induce differentiation of this same leukemic cell into macrophages [*Lotem and Sachs*, 1979]. These data indicate that actinomycin D, DMSO and TPA probably activate different genes in a given cell. It implies that they bind to and act on different sites of the DNA, the nature of which remains to be determined. We have mentioned in a previous chapter that depending on the concentration used, actinomycin D preferentially inhibits the synthesis of rRNA while that of other types of RNA is hardly affected under the same conditions. This indicates that actinomycin D might bind to certain specific sites of DNA as we have shown in recent studies [*Beljanski* et al., 1981a]. The specificity of substances used at low doses indicates that they are capable of recognizing the limits of one or more structural gene or operon. One tends to forget that DNA is not present in cells in the physicochemical form as described in the theoretical models where two DNA chains are held together by hydrogen bonds. Chromosomal DNA is in interaction not only with histones but also with different proteins, enzymes, amino acids, RNA, etc. Even when thoroughly purified, DNA contains peptides, amino acids, RNA fragments or some other related molecules [*Hillar and Przyjenski*, 1973; *Beljanski*, unpublished results]. The number and varieties of these substances may be responsible for the stability or instability of the DNA double helix. Activation of chromosomal genes by an endogenous or exogenous agent either requires an alteration in chromatin structure which makes the gene accessible to a given agent or else removal of histones [*Allfrey* et al., 1978]. When observations obtained in vitro and in vivo are combined, changes in the chromosomal level and even in the DNA itself in cell nuclei should consequently be detectable under the appropriate conditions. Thus, for example, 2-fluorene acetamide (2-FAA), a carcinogen, administered to rats or mice, results in the appearance of neoplastic nods in the liver. The DNA of the cells in these nods is in a 'destabilized form' [*Slifkin* et al., 1970]. Cells possessing such a DNA divide very rapidly compared to cells outside the nods. Nucleus enlargement observed in the presence of a carcinogen is also accompanied by accelerated cell division. This indicates that the agent used has set in motion the various steps of cell division. However, these observations only show that changes occur in the chromatin and DNA but do not indicate which genes are activated. For some authors, gene activation depends on DNA replication [*Harris*, 1973], while for others gene replication seems secondary to gene transcription [*Klevecz and Hsu*, 1964]. It appears that both activation and inactivation of genes take place in this process, and large-scale restriction of the synthesis of different mRNAs has been observed [*Grouse* et al., 1972; *Davidson,* 1968]. On the other hand, the expression of certain genes has to be amplified in order to give rise to a specialized cell population.

It has been suggested that molecules which bind to the single chains of DNA may act as 'gene derepressors' while those which preferentially bind to double-stranded mDNA may have the task of preventing gene transcription [*Frenster,* 1965, 1976]. This hypothesis, although attractive, cannot be generalized for three reasons:

(1) A given substance, actinomycin D for example, which binds either to the surface of double-stranded DNA or intercalates between two chains, has two different effects: small concentrations induce the liberation of specific information (differentiation), while high concentrations block transcription. Actinomycin D is known to bind only to double-stranded DNA.

(2) Some alkaloids of the β-carboline class used at low or high concentrations bind specifically to initiation sites for replication of cancer DNA, thus preventing DNA synthesis [*Beljanski and Beljanski,* 1982].

(3) Particular RNA fragments which bind to double-stranded DNA from spleen and bone marrow strongly stimulate DNA synthesis in vitro as well as genesis of platelets and leukocytes in vivo and even differentiation [*Beljanski* et al., 1978c]. On the other hand, the same RNA fragments bind to double-stranded DNA of various cancer tissues but do not stimulate their DNA synthesis in vitro nor the multiplication of cancer cells in mice. They do not enhance pathogenic leukocyte genesis. RNA fragments bind in both cases to double-stranded DNA. These observations support the view that the quantitative and qualitative effects of a substance which binds to DNA [*Gabbarro-Arpa* et al., 1979] might lead to the differential expression of certain DNA segments with accurate transcription.

Model for the Molecular Mechanism of Gene Activation or Inactivation; General Considerations

During differentiation and development the cells of a multicellular organism are differentially multiplied in order to organize specific tissues. Experimental data show that genetic properties of an organism are present in the egg and that both activation and/or inactivation of genes takes place during differentiation and development. In addition, differentiation requires the amplification in the expression of some genes. On the other hand, various experimental data have shown that the genomes of normal and tumor cells remain genetically equivalent. Thus, plant teratoma cells, grown under the appropriate conditions, give rise to a normal plant [*Braun*, 1972]. Mice teratocarcinoma cells, injected into the cavity of slightly older (blastocyst-stage) embryos, participate in the formation of many tumor-free mosaic mice [*Mintz and Illmensee*, 1975]. In these mice the coexisting tumor-lineage cells are no more malignant. According to these authors, the teratocarcinoma cells, although karyotypically entirely normal, came about through 'aberrant gene expression' and not through mutation. The concept of apparent irreversibility of the tumorous state cannot be applied to all types of cancer cells.

In cells of an organism there are functional and apparently nonfunctional or poorly functional genes. Gene activation (transcription) or inactivation (inhibition of transcription) requires molecules which allow or prevent binding of RNA polymerase to DNA as well as its full activity. This implies local opening or closing of the DNA helix. It should be remembered that the classical model of gene repression or derepression is a very simple model which cannot account for apparent irreversible gene activation or inactivation.

Basis for the Model

The basis for the model we propose is supported by data concerning specific gene activation 'in vivo', cell differentiation (mainly cells or organs in culture), and our recent results obtained by studying the effect of various compounds on in vitro behavior of DNA from cancer cells and healthy tissues, respectively. (Cancer cells have several characteristics of embryonic cells, while the 'healthy' DNA was from adult tissue cells.) These investigations showed that various endogenous and/or exogenous molecules induce in vitro local DNA helix opening with high efficiency in cancer DNA, but only slightly in DNA from healthy tissues. These physicochemical changes in DNA are observed particularly when the molecules which stimulate DNA synthesis in vitro are present in relatively low concentrations [*Beljanski* et al., 1981a]. The correlation between local DNA strand separation in vitro and cancer cell multiplication in vivo and the fact that these same agents are capable of inducing differentiation in several types of cancer cells (chapt. 4, 5) strongly suggest that they essentially act by inducing local DNA helix openings. Since this process does not occur spontaneously at physiological pH and temperatures, participation of enzymes, proteins, peptides, hormones, carcinogens (naturally produced or exogenously introduced), amino acids, oligoribonucleotides or similar agents are required. It is generally assumed that a given contiguous region of double-stranded DNA corresponds to a gene and that only one strand is transcribed into mRNA. Although some carcinogens or steroids appear to know how to induce the action of a transcriptional unit, it remains difficult to define the limits of selective transcription. This is well illustrated by findings that there are long chains of precursor RNAs which have to be cleaved by splicing enzymes in order to produce much smaller but mature mRNA. The recent discovery of 'mosaic mRNA' in eukaryotes indicates that the length of a transcribed segment of DNA is much larger than the length of the mature 'mosaic mRNA'. One transcription unit apparently can correspond to many polypeptide chains of related or differing functions [*Gilbert,* 1978]. In vitro experiments have shown that RNA polymerase prefers single- to double-stranded DNA. This requires in vivo freeing of DNA from histones or other molecules that stabilize the DNA helix. Molecules which act as gene activators have to reach the DNA. One fraction might interact with histones, RNA or other molecules bound to DNA, in order to remove them from strategic points, while another fraction may directly produce local DNA strand separation, as we have shown recently using intact and purified native DNA from cancerous and normal cells. The model we propose (fig. 29) may account for these observations and theoretical considerations. As shown on figure 29, a gene activator when used at low concentrations binds to only one strand in a given region of DNA freed from histones. When the first hydrogen bond linkage in the DNA helix is broken, the second and later ones follow, leading thus to a local separation of DNA strands. One strand is transcribed to mRNA, while the other one which binds some molecules may transiently undergo physicochemical changes for a period of time. Oligoribonucleotides, which act as primers for replication of only one DNA strand (lagging strand), may protect that strand against transcription. This would agree with the view that gene replication has to be secondary to transcription [*Klevecz and Hsu,* 1964]. When a newly activated gene is surrounded by favorable environmental conditions, it may remain functional – if the conformational structure of DNA in the neighbourhood of it has changed. In addition to histones, various other molecules may interact with these portions of DNA in order to maintain the new gene in a functional state. This could lead to the organization of some nucleosome cores in connection with newly activated genes, a process which depends on diffusible molecules and not on genetic interference [*Sperling* et al., 1980]. When molecules which interact with DNA are used at high concentrations

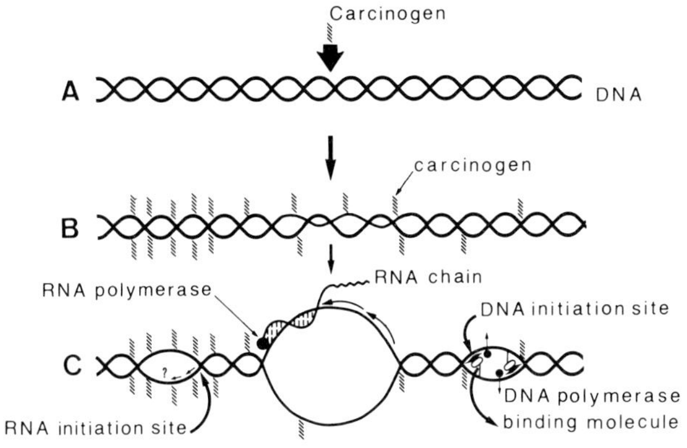

Fig. 29. Schematic representation of DNA chain opening and closing in the presence of substances binding to DNA (intercalation or not) [*Beljanski*, unpublished].

they bind at any place of both strands (fig. 29), thus preventing transcription and/or replication. On theoretical grounds a gene activator should not bind to initiation sites of the DNA transcriptional strand or to DNA replication sites. These sites have to be free in order to accept the appropriate polymerase. This view appears to be supported by our in vitro experiments concerning DNA helix opening and closing. In fact, we have shown that some alkaloids of the β-carboline class bind specifically to initiation sites for DNA replication, thus preventing DNA synthesis. No effect on chain elongation was observed. These alkaloids preferentially bind to DNA from cancer cells because this DNA has more unpaired regions than DNA from healthy cells. It is conceivable that closing of unpaired chains of cancer DNA to a threshold level which corresponds to that in normal DNA may lead to the conversion of cancer cells to normals. The unpaired regions probably contain the initiation sites in greater number since cancer DNA is a much better template for DNA synthesis than DNA from healthy cells. When carcinogens (or steroid hormones) liberate the initiation sites on DNA in vitro, leading to the increase of DNA synthesis, alkaloids introduced into the incubation mixture during the reaction will rapidly prevent DNA or RNA synthesis. They bind more rapidly to DNA (initiation sites) than either DNA or RNA polymerase. These data strongly suggest that carcinogens or other gene activators may bind near to initiation sites of DNA while alkaloids induce the closing of DNA chains, resulting in gene inactivation. These compounds bind to a specific set of DNA segments rich in G and A bases. Non-covalent interaction may occur which affects some phosphate groups of DNA. The molecular mechanism of gene activation and inactivation proposed here allows to interpret in a coherent manner many data concerning gene activation/inactivation, cell transformation and differentiation which all take place in the presence of various endogenous or exogenous molecules. This model may also throw a new light on the mechanism which may operate in the evolution of new animal and plant species. It harmonizes with a basic concept on the mechanism of evolution [*Wolsky and de Issekutz-Wolsky,* 1982].

Summary

We propose here a basic molecular mechanism of gene activation and inactivation, both of which may be induced by various endogenous or exogenous molecules. Many data show that selective gene activation requires factors which permit the transcription of genes with accurate output. Compounds exist which may induce the activation of certain genes without affecting others. Some compounds preferentially bind to unpaired strands in DNA, while others prefer the double-helical structure. There are also specific binding sites on DNA molecules for different compounds which, when attached to the site, liberate the initiation sites for replication, while other compounds specifically interfere only with initiation sites, thus preventing DNA and RNA synthesis and, consequently, cell division. This was demonstrated in particular with DNA from cancer cells. Depending on their nature and concentrations, DMSO, steroid hormones, carcinogens, RNA fragments, phorbol esters, and other substances may induce in vitro local opening (or closing) of DNA chains, thus enabling or preventing DNA replication and/or transcription. A remarkable fact is that those agents which induce DNA strand separation may, under specified conditions, induce cell differentiation or division. There is a correlation between DNA synthesis and strand separation in vitro, and multiplication of cancerous animal and plant cells in vivo, as was demonstrated by using certain carcinogens or guanidines. On the other hand, there are other substances, alkaloids of the β-carboline class and various proteins, peptides and RNA fragments, which prevent DNA strand separation and, consequently, DNA and RNA synthesis. Physicochemical changes of DNA and its chain opening or closing which occur in the presence of the substances mentioned appear to be of much greater importance than possible alkylation of some bases in DNA or mRNA as discussed above.

Final Discussion and Recapitulation

Cell differentiation and transformation induced by a great number of various environmental agents (excluding exogenous DNAs) is a fundamental biological process which requires rational explanation on the molecular level. One of the key problems is the basic mechanism of gene activation and inactivation related to DNA replication, which is regulated biochemically, physiologically and morphologically. These are complex events because DNA in a cell is in constant interaction with many endogenous and exogenous molecules. This situation is far different from that created in in vitro studies using isolated DNA. Replication and/or transcription requires strand separation, and this does not occur spontaneously at physiological temperatures.

DNA-dependent DNA polymerases do not recognize the individual initiation points which are present in a very large number of DNA molecules. Moreover, DNA replication requires the participation of various molecules: untwisting and unwinding enzymes, RNA primers, nucleases, divalent cations, nucleotides, etc. Discontinuous DNA synthesis starts at the replication fork and moves progressively away from the initiation site. Thus, all those substances which act either on initiation points or on elongation interfere with DNA synthesis. Active short-chain RNA primers bind to initiation points on intact DNA (single-chain), thus providing the 3′-OH group necessary for initiation of a new chain, although there is no certainty that both strands require RNA primers. Different initiation points on DNA seem to require different RNA primers. RNA primers might be synthesized by RNA polymerase or provided by RNases which act on multiple extra copies of various RNAs, among them rRNAs. These RNAs disappear as soon as cell growth starts. In addition, some short RNA fragments may prevent DNA replication and transcription and, consequently, also cell division (the case of plant cancer cells).

In contrast to DNA polymerase, RNA polymerases appear to know how to open DNA strands and to initiate the transcription. However, this process also requires many other factors. Under the normal physiological conditions of cell division there is a correlation between replication, transcription and translation. In nondividing cells there must be an interdependent correlation between transcription of certain genes (while others may remain silent) and translation.

The extraordinary abundance and variety of carcinogens and chemicals (antibiotics, antimitotics, etc.) in our environment raises the question of their active participation in DNA replication and transcription and subsequent cell multiplication and differentiation. Many examples cited above demonstrate that these agents, as well as many endogenously made substances (steroids, peptides, amino acids, etc.), interact with DNA (either single- or double-stranded), thus modifying its physicochemical structure with no necessity to covalently attach to DNA. Whereas some of them (the nature and concentration are important) destabilize the double-helical structure through their binding to DNA, others stabilize it. To give the advantage either to DNA polymerase or RNA polymerase, the cell must possess a series of molecules, capable of chosing the initiation and termination points in coordination with the appropriate polymerase. During gene activation, RNA polymerase should be allowed to transcribe DNA for at least a certain period of time. DNA polymerase is silent at this time in this region.

Beside electrophilic intermediates which reach nuclear DNA, carcinogens and many other compounds greatly influence the geometry of the 'DNA-compound' complex, leading to the opening or closing of DNA chains in certain areas. These phenomena were demonstrated in particular with DNAs from cancerous mammalian and plant cells. Since cancer DNAs are already partly destabilized, they respond more actively to various substances compared with DNA from healthy cells. Low concentrations of carcinogens induce in vitro local cancer DNA strand separation, accelerate DNA synthesis in vitro and enhance cancer cell multiplication in vivo. Under the same conditions, DNAs from healthy cells respond to them with lower efficiency. Carcinogens and/or hormones need to act on DNA persistently in healthy cells in order to destabilize the DNA chains, thus facilitating the attachment of molecules. This leads to the appearance of cancer cells. In the case of plant cells, the changes of the normal physicochemical state of DNA, obtained, for example, by the persistent use of plant hormones at given concentrations in the culture medium of healthy cells, leads to the appearance of 'anergic cells', i.e., cells with a destabilized DNA, as is the case with DNA from crown-gall tissues. Guanidines, abundantly synthesized by cancerous plant cells, induce in vitro DNA strand separation of cancer DNA and DNA from anergic tissues. Guanidines, as gene products, contribute to maintain the destabilized DNA in plant cancer cells. They stimulate cancer cell multiplication in vivo.

Although DNA strand separation can be initiated by RNA polymerase itself, hormones, carcinogens and some peptides, acting as gene activators, will further destabilize cancer DNAs and, in addition, maintain the unpaired segments of DNA in the length of a gene, an operon or even a larger segment. This gives rise to an increase in mRNA output, particularly in those DNA segments which were previously poorly transcribed. DNA, the ultimate target of these agents, contains sites which are specific for a certain agent. The nature of these sites is unknown. The appearance of specific mosaic mRNAs (through the splicing of precursor RNAs) and increased protein synthesis in the presence of carcinogens has been more frequently observed with cancer cells than with normal ones, probably on account of the destabilized structure of DNAs in the former, which enhances their reactivity. Carcinogens may also induce in vitro cancer cell differentiation at low concentrations, while at high doses they provoke degeneration of both normal and cancerous mammalian and plant cells. According to their nature they might differentiate a certain cell type into cells exhibiting different properties.

There are molecules, such as alkaloids of the β-carboline group, which specifically bind to the initiation points of destabilized DNA and thus prevent enzymes, carcinogens and hormones from separating DNA strands. These alkaloids do not allow DNA or RNA synthesis and suppress the binding of RNA primers to the initiation points of such DNAs. They have little effect on stabilized DNA, i.e., DNA from healthy cells.

Physicochemical changes of DNA, including local strand separation, also appear to be connected with the morphological and biochemical changes which exogenous RNAs may induce in various biological systems such as eukaryotic cells or tissues, fungi and bacteria. In most cases the mechanism for cell transformation or differentiation induced by specific mRNAs has not been elucidated on the biochemical level, although the experimental data seem to be convincing. It has been suggested, however, that reverse transcriptase could play an important role in this process by transcribing RNAs into DNA. The transcript may, in turn, become integrated into the genome of recipient cells. This process would require DNA strand separation for the integration of newly made DNA. DNA strand opening is required for the binding of specific small-size RNA fragments or oligoribonucleotides which accelerate the DNA replication of hematopoietic cells. They induce the multiplication and differentiation of stem cells and of their intermediates in animals and humans. This leads to normal levels of white blood cells and platelets. Other types of RNA fragments bind to destabilized plant cancer DNA, thus inhibiting the multiplication of cancer cells in vivo. It should be noted that large-size mRNA used for transformation or differentiation might produce oligoribonucleotides (through RNase activity) which, when rich in purines, will activate translation, but when rich in pyrimidine bases will act as inhibitors of it. In this way, they either promote or prevent interdependent transcription and thus contribute to cell economy. Since many oligoribonucleotides can be produced by RNases, the role of these enzymes appears to be of crucial importance during cell transformation, differentiation or dedifferentiation. The increase or decrease in RNase synthesis is the tool for the separation or closing of DNA strands which might be induced by carcinogens, tumor promoters, RNA fragments or certain specific alkaloids.

Many experimental data have shown that carcinogenesis and differentiation of cultured cells in vitro, i.e., the two opposite poles of cellular physiology, may be induced and controlled by various endogenous or exogenous substances. This happens through local, limited DNA strand opening or closing. The fact that a cancer cell may differentiate in the presence of DMSO, tumor promoters or carcinogens into a normal cell strongly indicates that the cancerous state of a cell is not caused by mutations, i.e., changes in the structure of the genic DNA itself.

The model for gene activation or inactivation which we propose here is mainly based on data obtained from comparing the activity of DNA from cells of adult organisms (differentiated cells) with that from tumor cells of animals or plants (cells which possess some characteristics of embryonic cells).

Could this model be used to explain the process of cell differentiation and the appearance of cancer cells?

Differentiated cells in different tissues of an organism possess the same genetic material. However, they exhibit different biochemical and morphological functions due to the fact that some genes are activated and functional while others are silent (repressed). Embryologists have often demonstrated that extrachromosomal molecules affect DNA. The establishment of a program in the cytoplasm will largely determine the DNA activity ('feedback mechanism'). As we have shown in this monograph, various substances may induce or prevent the normal progress of transcription. Also some segments of DNA are poorly transcribed. These events correspond to the extent of modulation of the physicochemical structure of DNA, i.e., local opening or closing of DNA strands. This modulation is compatible with progression of differentiation in pluricellular organisms. The classical theory of cell differentiation is essentially based on the concept of differential activation of genes in the course of development. In other words, the process of differentiation implies the appearance of differences in the activity of genes. The effects of multiple cytoplasmic or exogenous molecules in the activation of genes and differentiation of cells cultured in vitro fit well into our proposed model for modulation of gene activation or inactivation. On the basis of this model it is conceivable that the persistent and cumulative effects of either carcinogens or hormones or pH value on healthy cells lead to the appearance of cancer cells in which the DNA structure is destabilized. Activation of one gene in a normally differentiated cell may contribute to the repression of some other genes, thus returning the cell to the embryonic state (dedifferentiation). If environmental conditions are in favor of this cell, newly activated genes and their products may take over the role of the repressed genes, and another type of cell differentiates (transformation). In contrast, an exogenous agent may, according to our model, convert cancer cells to normal ones, as shown by several examples (DMSO, TPA, actinomycin D) discussed in this monograph. Stabilized DNA structure in normal cells and destabilized DNA in cancer cells from animals and plants are under the control of gene products. Each DNA responds differently to multiple endogenous or exogenous molecules which impose local chain opening or closing and, consequently, determine the state of the cell.

Postscript
New Data Supporting the Biochemical Model Described in this Monograph

Recently obtained results [*Neubort, S.; Liebeskind, D.; Mendez, F.; Elequin, F.; Hsu, K.S.; Bases, R.:* Morphological transformation, DNA strand separation and antinucleoside. Immunoreactivity following exposure of cells to intercalating drugs. Molec. Pharmacol. *21:* 739–743, 1982] on DNA strand separation in eukaryotic cells are in agreement with our published observations and support the biochemical model for gene activation and inactivation I presented in the present monograph. In fact, using quinacrine and proflavine, two intercalating agents and mutagens, these authors have shown that both induce increases in single-stranded DNA detected in the nuclei of mouse BALB C 3T3 1-13 cells. These compounds also induce rapid morphological transformation of mouse cells, although no covalent linkage between these compounds and cell components was detected.

References

Aaron-da Cunha, M.I.; Kurkdjian, A.; Le Goff, L.: Nature tumorale d'une hyperplasie obtenue expérimentalement. C.r. Séanc. Soc. Biol. *169:* suppl. 3, pp. 755–760 (1975).

Aaronson, S.A.; Parks, W.P.; Scolnick, E.M.; Todaro, G.J.: Antibody to the RNA-dependent DNA polymerase of mammalian-C type RNA tumor viruses. Proc. natn. Acad. Sci. USA *68:* 920–924 (1971).

Abdel-Monem, M.; Hoffman-Berling, H.: DNA unwinding enzymes. Trends biochem. Sci., pp. 128–130 (May 1980).

Abelev, G.I.; Perova, S.D.; Khramkova, I.N.; Postnikova, Z.A.; Irlin, I.S.: Production of embryonal α-globulin by transplantable mouse hepatomas. Transplantation *1:* 147–180 (1963).

Abelson, H.T.; Robstein, L.S.: Influence of Prednisolone on Moloney leukemogenic viruses in BALB/C mice. Cancer Res. *30:* 2208–2212 (1970).

Abrell, J.W.; Smith, R.G.; Robert, M.S.; Gallo, R.C.: DNA polymerase from RNA tumor viruses and human cells: inhibition by polyuridylic acid. Science *177:* 1111–1114 (1972).

Adolf, G.R.; Swetly, P.; Ludwig, H.: Tumor-promoting compound (TPA) enhances interferon on production in chronic lymphatic leukemia cells. Br. J. Haemat. *48:* 343–384 (1981).

Akagi, K.; Yamanoka, M.; Murai, K.; Niho, Y.; Omae, T.: Serum acid ribonuclease in myelogenous leukemia. Cancer Res. *38:* 2168–2173 (1978).

Akinrimisi, E.; Bonner, J.; Ts'o, P.O.P.: Binding of basic proteins to DNA. J. molec. Biol. *11:* 128–136 (1965).

Albanese, A.A.; Lorenze, E.J.; Orto, L.A.; Wein, E.H.; Zavattaro, D.N.; De Carlo, R.: Nutritional and metabolic effects of some newer steroids. VI. Serum ribonuclease. N.Y. St. J. Med., pp. 1595–1600 (June 15, 1972).

Allfrey, V.G.: Functional aspects of DNA associated proteins in histones and nucleohistones; in Phillips, Histone and nucleohistones, pp. 241–294 (Plenum Press, New York 1971).

Allfrey, V.G.; Boffa, L.C.; Vidalt, G.: Changes in composition during chemical carcinogenesis. Association of tumor-specific DNA-binding proteins with DNase I-sensitive region of chromatin and observation on the effects of histone acetylation on chromatin structure; in Ahmad, Russel, Schultz, Werner, Miami Winter Symposia (Differentiation and Development), vol. 15, pp. 261–294 (Academic Press, New York 1978).

Amaldi, F.; Carnevali, F.; Leoni, L.; Mariotti, D.: Replicon origins in Chinese hamster cell DNA. Expl Cell Res. *74:* 367–374 (1972).

Anderson, P.; Bauer, W.: Supercoiling in closed circular DNA: dependence upon ion type and concentration. Biochemistry *17:* 594–601 (1978).

Anthony, D.D.; Zeszotek, E.; Goldthwait, D.A.: Initiation by the DNA-dependent RNA polymerase. Proc. natn. Acad. Sci. USA *56:* 1026–1033 (1966).

Aposhian, H.V.; Kornberg, A.: Enzymatic synthesis of deoxyribonucleic acid. The polymerase formed after T_2 bacteriophage infection of *E. coli.* A new enzyme. J. biol. Chem. *237:* 519–525 (1962).

Arcos, J.C.; Argus, M.S.: Chemical induction of cancer, vol. II, A and B (Academic Press, New York 1974).

August, J.; Ortiz, P.; Hurwitz, J.: Ribonucleic acid dependent ribonucleotide incorporation. I. Purification and properties of the enzyme. J. biol. Chem. *237:* 3786–3793 (1962).

Axel, R.: Cleavage of DNA in nuclei and chromatin with staphylococcal nuclease. Biochemistry *14:* 2921–2925 (1975).

Baird, W.M.; Sedgwick, J.A.; Boutwell, R.K.: Effects of phorbol and four diesters on the incorporation of tritiated precursors into DNA, RNA and protein in mouse epidermis. Cancer Res. *31:* 1434–1439 (1971).

Baker, H.W.G.; Burger, H.G.; De Kretser, D.M.; Hudson, B.; Staffon, W.G.: Effects of synthetic oral estrogens in normal men and patients with prostatic carcinoma. Lack of gonadotropin suppression by chlorotrianisene. Clin. Endocrinol. *2:* 297–306 (1974).

Ball, L.A.; White, C.N.: Nuclease activation by double-stranded RNA and by 2′,5′-oligoadenylate in extracts of interferon-treated chick cells. Virology *93:* 348–356 (1979).

Barghava, P.; Shanmugam, G.: Uptake of nucleic acids by mammalian cells. Prog. nucleic Acid Res. *11:* 104–192 (1971).

Barker, G.R.; Bray, C.B.; Walter, T.J.: The development of ribonuclease and acid phosphatase during germination of *Pisum arvense.* Biochem. J. *142:* 211–219 (1974).

Barker, K.L.; Warren, J.C.: Template capacity of uterine chromatin: control by estradiol. Proc. natn. Acad. Sci. USA *56:* 1298–1302 (1966).

Barra, R.; Hicks, H.; Koch, M.R.; Lea, M.A.: Stimulatory effect of dimethylsulfoxide on (^{3}H)-thymidine incorporation into DNA in Novikoff hepatoma cells. Int. J. Biochem. *916:* 389–394 (1978).

Barrell, B.G.; Air, G.M.; Hutchison, C.A., III: Overlapping genes in Bacteriophage ΦX174. Nature, Lond. *264:* 34–41 (1976a).

Barrell, B.G.; His, G.M.; Hutchison, C.A., III: Bacteriophage ΦX174 genes D and F are translated from the same DNA sequence but in different reading frames. Nature, Lond. *264:* 34–41 (1976b).

Barry, J.; Gorski, J.: Uterine ribonucleic acid polymerase. Effect of estrogen on nucleotide incorporation into 3′ chain termini. Biochemistry *10:* 2384–2390 (1971).

Barth, L.G.: Neural differentiation without organizer. J. exp. Zool. *87:* 371–383 (1941).

Baserga, R.: Multiplication and division in animal cells (Decker, New York 1976).

Baumgartner, B.; Harris, N.: Regulation of reserve protein metabolism in the cotyledons of mung bean seedlings. Proc. natn. Acad. Sci. USA *73:* 3168–3172 (1976).

Baumgartner, B.; Matile, P.: Immunocytochemical localization of acid ribonuclease in morning glory tissue. Biochem. Physiol. Pflanzen *170:* 279–285 (1976).

Bautz, E.K.F.; Bautz, F.A.: Initiation of RNA synthesis: the function of σ in the binding of RNA polymerase to promoter sites. Nature, Lond. *226:* 1219–1222 (1970).

Bear, M.P.; Schneider, F.H.: The effects of actinomycin D and cordycepin on neurite formation and acetylcholinesterase activity in mouse neuroblastoma cells. Expl Cell Res. *123:* 301–309 (1979).

Beljanski, M.: Synthèse in vitro de l'ADN sur une matrice d'ARN par une transcriptase d'*E. coli.* C. r. hebd. Séanc. Acad. Sci., Paris, sér. D *274:* 2801–2803 (1972).

Beljanski, M.: ARN-amorceurs riches en nucleotides G et A indispensables à la réplication in vitro de l'ADN des phages ΦX174 et lambda. C. r. hebd. Séanc. Acad. Sci., Paris, sér. D *280:* 1189–1192 (1975).

Beljanski, M.: Oncotest: a DNA assay system for the screening of carcinogenic substances. IRCS med. Sci. *7:* 476 (1979).

Beljanski, M.; Aaron-da Cunha, M.I.: Particular small size RNA and RNA-fragments from different origins as tumor inducing agents in *Datura stramonium.* Mol. Biol. Rep. *2:* 497–506 (1976).

Beljanski, M.; Beljanski, M.: Synthese dans *Escherichia coli* des ARN dont la structure primaire diffère totalement de celle de l'ADN. C. r. hebd. Séanc. Acad. Sci., Paris, sér. D *267:* 1058–1060 (1968).

Beljanski, M.; Beljanski, M.: RNA-bound reverse transcriptase in *Escherichia coli* and *in vitro* synthesis of a complementary DNA. Biochem. Genet. *12:* 163–180 (1974).

Beljanski, M.; Beljanski, M.S.: Selective inhibition of cancer DNA *in vitro* synthesis by alkaloids of β-carboline class. Expl Cell Biol. *50:* 78–87 (1982).

Beljanski, M.; Beljanski, M.S.; Bourgarel, P.: ARN-transformants porteurs de caractères héréditaires chez *Escherichia coli* showdomycino-resistant. C. r. hebd. Séanc. Acad. Sci., Paris, sér. D *272:* 2107–2111 (1971a).

Beljanski, M.; Beljanski, M.S.; Manigault, P.; Bourgarel, P.: Transformation of *Agrobacterium tumefaciens* into a non-oncogenic species by an *Escherichia coli* RNA. Proc. natn. Acad. Sci. USA *69:* 191–195 (1972a).

Beljanski, M.; Beljanski, M.S.; Plawecki, M.; Bourgarel, P.: ARN-fragments amorceurs nécessaires à la réplication *in vitro* des ADN. C. r. hebd. Séanc. Acad. Sci., Paris, sér. D *280:* 363–366 (1975).

Beljanski, M.; Bonissol, C.; Kona, P.: Transformation des cellules KB induite par la showdomycine. C. r. hebd. Séanc. Acad. Sci., Paris, sér. D *274:* 3116–3119 (1972b).

Beljanski, M.; Bourgarel, P.; Beljanski, M.: Showdomycine et biosynthèse d'ARN non complémentaire de l'ADN. Annls Inst. Pasteur, Paris, *118:* 253–276 (1970).

Beljanski, M.; Bourgarel, P.; Beljanski, M.: Drastic alteration of ribosomal RNA and ribosomal proteins in showdomycin-resistant *Escherichia coli.* Proc. natn. Acad. Sci. USA *68:* 491–495 (1971b).

Beljanski, M.; Bourgarel, P.; Beljanski, M.: Découpage des ARN ribosomiques d'*Escherichia coli* par la ribonuclease U_2 et transcription *in vitro* des ARN fragments en ADN complémentaire. C. r. hebd. Séanc. Acad. Sci., Paris, sér. D *280:* 1825–1828 (1978a).

Beljanski, M.; Bourgarel, P.; Beljanski, M.: Correlation between *in vitro* DNA synthesis, DNA strand separation and *in vivo* multiplication of cancer cells. Expl Cell Biol. *49:* 220–231 (1981a).

Beljanski, M.; Kurkdjian, A.; Manigault, P.: Relation entre l'activité d'une *l*-asparaginase et le pouvoir oncogène de différents «mutants» d'*Agrobacterium tumefaciens* (Smith et Town) Conn. C. r. hebd. Séanc. Acad. Sci., Paris, sér. D *274:* 3560–3563 (1972c).

Beljanski, M.; Le Goff, L.; Aaron-da Cunha: Special short dual-action RNA fragments both induce and inhibit Crown-gall tumors. Proc. 4th Conf. Plant Path. Bot. Angers, p. 207–220 (1978b).

Beljanski, M.; Le Goff, L.; Beljanski, M.: *In vitro* screening of carcinogens using DNA of His^- mutant of *Salmonella thyphimurium.* Expl Cell Biol. (*50:* 271–280 (1982a).

Beljanski, M.; Le Goff, L.; Faivre-Amiot, A.: Preventive and curative anticancer drug. Application to Crown-gall tumors. Acta horticult. *125:* 239–248 (1982b).

Beljanski, M.; Manigault, P.: Genetic transformation of bacteria by RNA and loss of oncogenic power properties of *Agrobacterium tumefaciens*. Transforming RNA as template for DNA synthesis; in Beers, Tilghman, Cellular modification and genetic transformation by exogenous nucleic acids, suppl. 2, pp. 81–97 (The Johns Hopkins University Press, Baltimore 1972).

Beljanski, M.; Manigault, P.; Beljanski, M.S.; Aaron-da Cunha, M.I.: Genetic transformation of *Agrobacterium tumefaciens* B_6 by RNA and nature of the tumor inducing principle. Proc. 1st. Int. Congr. IAMS, vol. 1, pp. 132–141 (1974).

Beljanski, M.; Plawecki, M.: Transforming RNA as a template directing RNA and DNA synthesis in bacteria; in Niu, Segal, The role of RNA in reproduction and development, pp. 203–223 (North Holland/American Elsevier, New York 1973).

Beljanski, M.; Plawecki, M.; Bourgarel, P.; Beljanski, M.S.: Nouvelles substances (RLB) actives dans la leucopoiese et la formation des plaquettes. Bull. Acad. natn. Méd. *162:* 475–481 (1978c).

Beljanski, M.; Plawecki, M.; Bourgarel, P.; Beljanski, M.S.: Short chain RNA fragments as promoters of leukocyte and platelet genesis in animals depleted by anti-cancer drugs; in Niu, Chuang, The role of RNA in development and reproduction, pp. 79–113 (Science Press, Beijing 1981b; Van Nostrand & Reinhold Co).

Bendar, T.W.; Linsmaier-Bendar, E.M.: Induction of cytokinin-independent tobacco tissues by substituted fluoren. Proc. natn. Acad. Sci. USA *68:* 1178–1179 (1971).

Benedict, W.F.; Baker, M.S.; Haron, L.; Choi, E.; Ames, B.N.: Mutagenicity of cancer chemotherapeutic agents in the *Salmonella*/microsome test. Cancer Res. *37:* 2209–2213 (1977).

Benjamin, W.; Levander, O.A.; Gellhorn, A.; Debellis, R.H.: An RNA-histone complex in mammalian cells: the isolation and characterization of a new RNA species. Proc. natn. Acad. Sci. USA *55:* 858–865 (1966).

Benz, E.W.; Reinberg, D.; Vicuna, R.; Hurwitz, J.: Initiation of DNA replication by the DNA G protein. J. biol. Chem. *255:* 1096–1106 (1980).

Berenblum, I.: The carcinogenic action of croton resin. Cancer Res. *1:* 44–48 (1941).

Berenblum, I.: Sequential aspects of chemical carcinogenesis of skin; in cancer, a comprehensive treatise, vol. 1, pp. 323–338 (Plenum Press, New York 1975).

Berns, A.; Salden, M.; Bogdanowsky, D.; Raymondjean, M.; Schapira, G.; Blomendel, H.: Non-specific stimulation of cell-free protein synthesis by a dialysable factor isolated from Reticulocyte Initiation Factors ('i-RNA'). Proc. natn. Acad. Sci. USA *72:* 714–718 (1975).

Bertazzoni, C.; Chieli, T.; Solcia, E.: Different incidence of breast carcinomas or fibroadenomas in daunomycin or adriamycin treated rats. Experiantia *27:* 1209–1210 (1971).

Bertazzoni, U.; Stefanini, M.; Pedrali-Noy, G.; Guilotto, E.; Nuzzo, F.; Falaschi, A.; Spadari, S.: Variation of DNA polymerases α and β during prolonged stimulation of human lymphocytes. Proc. natn. Acad. Sci. USA *73:* 785–789 (1976).

Bester, A.J.; Kennedy, D.S.; Heywood, S.M.: Two classes of translational control RNA: their role in the regulation of protein synthesis. Proc. natn. Acad. Sci. USA *72:* 1523–1527 (1975).

Bieri-Bonniot, F.; Joss, U.; Dierks-Venteing, C.: Stimulation of RNA polymerase I activity by 17β-estradiol-receptor complex on chick liver nucleolar chromatin. FEBS Lett. *81:* 91–96 (1977).

Bischoff, R.; Holtzer, H.J.: The effect of mitotic inhibitors on myogenesis in vitro. J. Cell Biol. *36:* 111–127 (1968).

Blair, D.G.; Sheratt, D.J.; Clewell, D.B.; Helinski, D.R.: Isolation of supercoiled colicinogenic factor E_1 DNA sensitive to ribonuclease and alkali. Proc. natn. Acad. Sci. USA *69:* 25–18 (1972).

Blau, H.M.; Epstein, J.: Manipulation of myogenesis in vitro: reversible inhibition by DMSO. Cell *17:* 95–108 (1979).
Blinkerd, P.E.; Toliver, A.P.: Association of RNA with discontinuous DNA replication in HeLa cells. Cytobios *10:* 221–233 (1974).
Blomberg, P.M.; Robbins, P.W.: Effect of proteases on activation of resting chick embryo fibroblasts and on cell surface proteins. Cell *6:* 137–147 (1975).
Bollum, F.J.: Calf thymus polymerase. J. Biol. Chem. *235:* 2399–2403 (1960).
Bonner, J.; Dahmus, M.E.; Fambrough, D.; Huang, R.C.; Marushige, K.; Tuan, D.Y.H.: The biology of isolated chromatin. Science *159:* 47–56 (1968).
Borsa, J.; Whitemore, G.F.: Studies relating to the mode of action of methotrexate. III. Inhibition of thymidylate synthetase in tissue culture cells and in cell free systems. Molec. Pharmacol. *5:* 318–321 (1969).
Borthwick, N.M.; Smelie, R.M.S.: Effects of estradiol-17β on the ribonucleic acid polymerases of immature rabbit uterus. Biochem. J. *147:* 91–101 (1975).
Bourgaux, P.; Bourgaux-Ramoisy, D.: Unwinding of replicating polyoma virus DNA. J. molec. Biol. *70:* 399–413 (1972).
Brachet, J.: La localisation des acides pentose-nucleiques dans les tissus animaux et les œufs d'amphibiens en voie de développement. Archs Biol., Paris *53:* 207–257 (1942).
Braun, A.C.: The relevance of plant tumor systems to an understanding of the basic cellular mechanisms underlying tumorigenesis. Prog. exp. Tumor Res., vol. 15, pp. 165–187 (1972).
Braun, A.C.; Wood, H.N.: On the inhibition of tumor inception in the Crown-gall disease with the use of ribonuclease A. Proc. natn. Acad. Sci. USA *56:* 1417–1422 (1966).
Brawerman, B.; Shapiro, R.; Szer, W.: Modification of *Escherichia coli* ribosomes and Coli phage MS2 RNA by bisulfite. Effects on ribosomal binding and protein synthesis. Nucl. Acids Res. *2:* 501–507 (1975).
Brewer, E.N.: DNA replication in *Physarum polycephalum.* J. molec. Biol. *68:* 401–412 (1972).
Briggs, R.; King, T.J.: Transplantation of living nuclei from bastula cells into enucleated frogs' eggs. Proc. natn. Acad. Sci. USA *38:* 455–463 (1952).
Britten, J.; Davidson, E.H.: Gene regulation for higher cells: a theory. Science *165:* 349–357 (1969).
Brody, E.; Diggelmann, H.; Geiduschek, E.P. Transcription of the bacteriophage template. Detailed comparison of *in vivo* and *in vitro* transcripts. Biochemistry *9:* 1289–1299 (1970).
Browerman, G.: Characteristics and significance of the polyadenylate sequence in mammalian messenger RNA. Prog. nucleic Acid Res. molec. Biol. *17:* 117–148 (1970).
Brown, D.D.; Blackler, A.W.: Gene amplification proceeds by a chromosome copy mechanism. J. molec. Biol. *63:* 75–83 (1972).
Brown, D.D.; Gurdon, J.B.: High-fidelity transcription of SSDNA injected into *Xenopus* oocytes. Proc. natn. Acad. Sci. USA *74:* 2064–2068 (1977).
Brown, D.D.; Littna, E.: Synthesis and accumulation of DNA-like RNA during embryogenesis of *Xenopus laevis.* J. molec. Biol. *20:* 81–94 (1966).
Brown, M.; Kiehn, D.: Protease effects on specific growth properties of normal and transformed baby hamster kidney cells. Proc. natn. Acad. Sci. USA *74:* 2874–2878 (1977).
Brown, R.D.; Tocchini-Valentini, G.P.: On the role of RNA in gene amplification. Proc. natn. Acad. Sci. USA *69:* 1746–1748 (1972).
Brown, W.M.; Shine, J.; Goodman, H.M.: Human mitochondrial DNA: analysis of 7S DNA from the origin of replication. Proc. natn. Acad. Sci. USA *75:* 735–739 (1978).

Brun, G.; Weissbach, A.: Initiation of HeLa DNA synthesis in a subnuclear system. Proc. natn. Acad. Sci. USA *75:* 5931–5935 (1978).

Bryant, J.A.: Biochemical aspects of DNA replication with particular reference to plants. Biol. Rev. *55:* 237–284 (1980).

Burgess, R.R.: RNA polymerase. A. Rev. Biochem. *40:* 711–740 (1971).

Busch, H.: The function of the 5′ cap of mRNA and nuclear species. Perspect Biol. Med. *19:* 549–567 (1976).

Byfield, J.E.; Lee, Y.V.; Tu, L.; Kulhanian, F.: Molecular interaction of the combined effects of bleomycin and X rays on mammalian cell survival. Cancer Res. *36:* 1136–1143 (1976).

Cairns, J.: Autoradiography of HeLa cell DNA. J. molec. Biol. *15:* 372–373 (1966).

Caplan, A.I.; Ordahl, C.P.: Irreversible gene repression model for control of development. Lessening of developmental potential may result from progressive repression of previously active genes. Science *201:* 120–130 (1978).

Castellano, T.J.; Schiffman, R.L.; Jacob, M.C.; Loeb, J.: Suppression of liver cell proliferation by glucocorticoid hormone: a comparison of normally growing and regenerating tissue in immature rat. Endocrinology *102:* 1107–1112 (1978).

Caujolles, F.M.E.; Caujolles, D.H.; Cross, S.B.; Calvett, M.M.: Limits of toxic and teratogenic tolerance to dimethylsulfoxide. Ann. N.Y. Acad. Sci. *141:* 110–125 (1967).

Center, M.S.: Induction of single-strand regions in nuclear DNA by Adriamycin. Biochem. biophys. Res. Commun. *89:* 1231–1238 (1979).

Chamberlin, M.J.: The selectivity of transcription. A. Rev. Biochem. *43:* 721–775 (1974).

Chamberlin, M.J.: Interaction of RNA polymerase with the DNA template; in Losick, Chamberlin, RNA polymerase, pp. 159–191 (Cold Spring Harbor Press, New York 1976).

Chamberlin, M.J.; Berg, P.: DNA directed synthesis of RNA by an enzyme from *E. coli.* Proc. natn. Acad. Sci. USA *48:* 81–85 (1964a).

Chamberlin, M.J.; Berg, P.: Mechanism of RNA polymerase action: characterization of the DNA-dependent synthesis of polyadenylic acid. J. molec. Biol. *93:* 219–227 (1964b).

Chambon, P.: Eukaryotic RNA polymerases. A. Rev. Biochem. *44:* 613–638 (1975).

Champoux, J.J.: Proteins that affect DNA conformation. A. Rev. Biochem. *47:* 449–479 (1978).

Chandebois, R.: Morphogénétique des animaux pluricellulaires (Maloine, Paris 1976).

Chandebois, R.: Cell sociology and the problem of automation in the development of pluricellular animals. Acta biotheor. *29:* 1–38 (1980).

Chang, L.M.S.: Low molecular weight DNA polymerase from calf thymus chromatin. II. Initiation and fidelity of homopolymer replication. J. biol. Chem. *248:* 6983–6992 (1973).

Chang, L.M.S.: The distributive nature of enzymatic DNA synthesis. J. molec. Biol. *93:* 219–235 (1975).

Chang, L.M.S.; Bollum, F.J.: Low molecular weight deoxyribonucleic acid polymerase in mammalian cells. J. biol. Chem. *246:* 5835–5837 (1971).

Chang, L.M.S.; Bollum, F.J.: A chemical model for transcriptional initiation of DNA replication. Biochem. biophys. Res. Commun. *46:* 1354–1360 (1972).

Chappell, J.; Van der Wilden, W.; Chrispeels, M.J.: The biosynthesis of ribonuclease and its accumulation in protein bodies in the codyledons of mung bean seedlings. Devl Biol. *76:* 115–125 (1980).

Chargaff, E.: Initiation of enzymatic synthesis of deoxyribonucleic acid by ribonucleic acid primers. Prog. nucleic Acid Res. mol. Biol. *18:* 1–24 (1977).

Chattoraj, D.; Stahl, W.: Evidence of RNA in D loops of intramolecular λ DNA. Proc. natn. Acad. Sci. USA *77:* 2153–2157 (1980).

Christman, J.K.; Price, P.; Pedrinan, L.; Acs, G.: Correlation between hypomethylation of DNA and expression of globin genes in Friend erythroleukemia cells. Eur. J. Biochem. *81:* 53–61 (1977).

Chuang, R.Y.; Chuang, L.F.: Increased frequency of initiation of RNA synthesis due to a protein factor from chicken myeloblastosis nuclei. Proc. natn. Acad. Sci. USA *72:* 2935–2939 (1975).

Church, R.B.; McCarthy, B.J.: Unstable nuclear RNA synthesis following estrogen stimulation. Biochim. biophys. Acta *199:* 103–114 (1970).

Clark, J.H.; Gorski, J.: Ontogeny of the estrogen receptor during early uterine development. Science *169:* 76–78 (1970).

Clark, J.H.; Paszko, Z.; Peck, E.J.: Nuclear binding and retention of the receptor estrogen complex: relation to the agonistic and antagonistic properties of estradiol. Endocrinology *100:* 91–96 (1977).

Clayton, R.B.; Kogura, J.; Kraemer, H.C.: Sexual differentiation of the brain: effects of testosterone on brain RNA metabolism in newborn female rats. Nature, Lond. *226:* 810–815 (1970).

Clemens, L.E.; Kleinsmith, L.J.: Specific binding of the oestradiol-receptor complex to DNA. Nature, Lond. *237:* 204–206 (1972).

Clever, U.: The ecdysone concentration dependent pattern of gene activity in the salivary gland chromosomes of *Tendipes tentans.* Devl Biol. *6:* 73–93 (1963).

Clever, U.; Karlson, P.: Induktion von Puffer-Veränderung in den Speicheldrüsenchromosomen von *Chironomus tentans* durch Ecdyson. Expl Cell. Res. *20:* 623–626 (1960).

Cohen, P.; Kidson, C.: Interactions of hormonal steroids with nucleic acids. I. A specific requirement for guanine. Proc. natn. Acad. Sci. USA *63:* 458–464 (1969).

Cohen, R.; Pacifici, M.; Rubinstein, N.; Biehl, J.; Holtzer, H.: Effect of a tumor promoter on myogenesis. Nature, Lond. *266:* 538–540 (1977).

Cole, R.S.; Levitan, D.; Sinden, R.R.: Removal of psoralen interstrand cross-links from DNA of *Escherichia coli:* mechanism and genetic control. J. molec. Biol. *103:* 39–59 (1976).

Colletta, G.; Frafonele, F.; Sandomenico, M.L.; Vecchio, G.: Enhanced expression of viral polypeptides and messenger RNA in dimethyl-sulfoxide and bromodeoxyuridine treated friend erythroleukemic cells. Expl Cell Res. *119:* 253–264 (1979).

Collins, S.J.; Rusceti, F.W.; Gallagher, R.E.; Gallo, R.C.: Terminal differentiation of human promyelocytic leukemia cells induced by dimethylsulfoxide and other polar compounds. Proc. natn. Acad. Sci. USA *75:* 2458–2462 (1978).

Congote, L.F.; Solomon, S.: Testosterone stimulation of a rapidly labeled low molecular weight RNA fraction in human hepatic erythroid cells in culture. Proc. natn. Acad. Sci. USA *72:* 523–527 (1975).

Connors, T.A.: Mechanism of action of 2-chloroethylamine derivatives, sulfar mustards, epoxides and aziridines; in Sartorelli, Johns, Antiplastic and immuno-suppressive agents, vol. II, pp. 18–34 (Springer, Berlin 1975).

Content, J.; Leblen, B.; De Clerg, E.: Differential effects of various double-stranded RNAs on protein synthesis in rabbit reticulocyte lysates. Biochemistry *17:* 88–94 (1978).

Costantini, F.D.; Britten, R.J.; Davidson, E.H.: Message sequences and short repetitive sequences are interspersed in sea urchin egg poly(A)t RNAs. Nature, Lond. *287:* 111–117 (1980).

Cowan, F.M.; Klein, D.L.; Armstrong, G.R.; Stylos, W.A.; Pearson, J.W.: Suppression of DNA synthesis in mitogen stimulated leukocytes with partial trypsin digests of *Staphylococcus aureus* protein A. Biomedicine *31:* 220–223 (1979).

Cox, R.F.; Haines, M.E.; Carey, N.H.C.: Modification of the template capacity of chick-oviduct chromatin for form-B RNA polymerase by estradiol. Eur. J. Biochem. *32:* 513–524 (1973).

Cox, R.P.: DNA methylase inhibition in vitro by N-methyl-N-nitro-N-nitroso-guanidine. Cancer Res. *40:* 61–63 (1980).

Craig, N.: The effect of inhibitors of RNA and DNA synthesis on protein synthesis and polysome levels in mouse L-cells. J. cell. Physiol. *82:* 133–150 (1973).

Crippa, M.; Davidson, E.H.; Mirsky, A.E.: Persistence in early amphibian embryos of informational RNAs from the lampbrush chromosome stage of oogenesis. Proc. natn. Acad. Sci. USA *57:* 885–892 (1967).

Cullis, C.A.; Davies, D.R.: Ribosomal DNA amounts in *Pisum sativum.* Genetics, N.Y. *81:* 485–492 (1975).

Curmingham, R.P.; Shibata, T.; Das Gupta, C.; Radding, C.M.: Single strands induce rec A protein to unwind duplex DNA for homologous pairing. Nature, Lond. *281:* 191–195 (1979).

Dahl, G.; Azarnia, R.; Werner, R.: Induction of cell-cell channel formation by m-RNA. Nature, Lond. *289:* 683–685 (1981).

Dahmus, M.E.; Bonner, J.: Increased template activity of liver chromatin, a result of hydrocortisone administration. Proc. natn. Acad. Sci. USA *54:* 1370–1375 (1965).

Darnell, J.E.: Implications of RNA. RNA splicing in evolution of eukaryotic cells. Science *202:* 1257–1261 (1978).

Darnell, J.; Wall, R.; Tushinski, R.: An adenylic acid-rich sequence in messenger RNA of HeLa cells and its possible relationship to reiterated sites in DNA. Proc. natn. Acad. Sci. USA *68:* 1321–1325 (1971).

Das, R.M.: Time-course of the mitotic response to estrogen in the epithelia and stroma of the mouse uterus. J. Endocr. *55:* 203–204 (1972).

David, J.C.; Vinson, D.; Bruner, F.: DNA-ligase changes and cell population in thymus of corticosteroid-treated chick embryo. Expl Cell Res. *130:* 137–146 (1980).

David, N.A.: The pharmacology of dimethylsulfoxide 6544. A. Rev. Pharmacol. *12:* 353–374 (1972).

Davidson, E.H.: Gene activity in early development, pp. 3–9 (Academic Press, New York, 1968).

Davidson, E.H.; Hough, B.R.: Genetic information in oocyte RNA. J. molec. Biol. *56:* 491–506 (1971).

Davis, R.W.; Hyman, R.W.: Physical locations of the in vitro RNA. Initiation site and termination sites of T7 M DNA. Cold Spring Harb. Symp. quant. Biol. *35:* 269–281 (1970).

De Groot, N.; Yuli, I.Y.; Czosnek, H.H.; Shiklosh, Y.; Hochberg, B.: Studies on the amino acid-incorporating activity of native rat liver rough membrane and that reconstituted *in vitro.* Biochem. J. *158:* 23–31 (1976).

de Issekutz-Wolsky, M.; Monteil, J.; Wolsky, A.: Tail regeneration in the newt *Triturus (Diemictylus) viridescens* under the influence of RNA preparations from adult tissues of the same species. Am. Zool. *6:* 614 (1966).

Delius, H.; Howe, C.; Kozinski, A.W.: Structure of the replicating DNA from bacteriophage T_4. Proc. natn. Acad. Sci. USA *68:* 3049–3053 (1971).

Delvin, R.B.; Emerson, C.P., Jr.: Coordinate accumulation of contractile proteins mRNAs during myoblast differentiation. Devl Biol. *69:* 202–216 (1979).

De Sombre, E.R.; Lyttle, C.R.: Steroid hormone regulations of uterine peroxidase activity. Adv. exp. Med. Biol. *117:* 157–171 (1979).

De Sombre, E.R.; Mohla, S.; Jensen, E.V.: Estrogen-independent activation of the receptor protein of calf uterine cytosol. Biochem. biophys. Res. Commun. *48:* 1601–1608 (1972).

De Sombre, E.R.; Mohla, S.; Jensen, E.V.: Receptor transformation. The key to estrogen action. J. Steroid Biochem. *6:* 469–473 (1975).

Desphande, A.K.; Niu, L.C.; Niu, M.C.: Requirement of informational molecules in heart formation; in Niu, Segal, The role of RNA in reproduction and development, pp. 229–246 (North Holland/American Elsevier, New York 1973).

Desphande, A.K.; Siddiqui, M.A.Q.: Acetylcholinesterase differentiation during myogenesis in early chick embryonic cells caused by inducer RNA. Differentiation *10:* 133–137 (1978).

De Young, L.M.; Argyris, T.S.; Gordon, G.B.: Epidermal ribosome accumulation during two-stage skin tumorogenesis. Cancer Res. *37:* 388–393 (1977).

Dhar, R.; Subramanian, R.; Pan, J.; Weissman, S.M.: Nucleotide sequence of a fragment of SV40 DNA that contains the origin of DNA replication and specifices the 5′ ends of 'early' and 'late' viral RNA. IV. Localization of the SV40 DNA complementary to the 5′ ends of viral mRNA. J. biol. Chem. *252:* 368–376 (1977).

Dicker, P.; Rozengurt, E.: Stimulation of DNA synthesis by tumour promoter and pure mitogenic factors. Nature, Lond. *276:* 723–726 (1978).

Dienstman, S.R.; Holtzer, H.: Myogenesis: a cell lineage interpretation; in Reinert, Holtzer, Cell cycle and cell differentiation, vol. 7, pp. 1–25 (Springer, Berlin 1975).

Dimarco, A.: Adriamycin (NSC 123127): mode and mechanism of action. Cancer Chemother. Rep. (part 3) *6:* 91–106 (1975).

Dimarco, A.; Zumino, F.; Silvestrini, R.; Garubarucci, C.; Gambetta, R.A.: Interaction of some daunomycin derivatives with deoxyribonucleic acid and their biological activity. Biochem. Pharmac. *20:* 1323–1328 (1971).

Di Mauro, E.; Snyder, L.; Marino, P.; Lamberti, A.; Coppo, A.; Tocchini-Valentini, G.P.: Rifampicin sensitivity of the components of DNA-dependent RNA polymerase. Nature, Lond. *222:* 533–537 (1969).

Dressler, D.: The recent excitement in the DNA growing point problem. A. Rev. Microbiol. *29:* 548–559 (1975).

Durrant, A.: The environmental induction of heritable change in Linum. Heredity *17:* 27–61 (1962).

Dutta, S.K.; Beljanski, M.; Bourgarel, P.: Endogenous RNA-bound RNA-dependent DNA polymerase activity in *Neurospora crassa.* Expl Mycol. *1:* 173–182 (1977).

Dutta, S.K.; Mukhopadhyay, D.K.; Bhattacharyya, J.: RNase-sensitive DNA polymerase activity in cell fractions and mutants of *Neurospora crassa.* Biochem. Genet. *18:* 743–753 (1980).

Dynan, W.S.; Burgess, R.R.: In vitro transcription by wheat germ RNA polymerase II. J. biol. Chem. *256:* 5866–5873 (1981).

Edelman, I.S.; Bogerach, R.; Poster, G.A.: The mechanism of action of aldosterone on sodium transport. The role of protein synthesis. Proc. natn. Acad. Sci. USA *50:* 1169–1177 (1963).

Edward, D.P.; Murthy, S.R.; Mc Guire, W.L.: Effects of estrogen and antiestrogen on DNA polymerase in human breast cancer. Cancer Res. *40:* 1722–1726 (1980).

Efstratiadis, A.; Kafatos, F.C.; Maniatis, T.: The primary structure of rabbit globin mRNA as determined from cloned DNA. Cell *10:* 571–585 (1977).

El-Sewedy. S.M.; El-Bassiouni, E.A.; Assar, S.T.: Effect of some steroids on bovine pancreatic ribonuclease activity in vitro. Biochem. Pharmac. *27:* 1831–1832 (1978).

Engel, J.D.; Hippel, P.H. von: Effects of methylation on the stability of nucleic acid conformations. J. biol. Chem. *253:* 927–934 (1978).

Epifanova, O.I.: Mitotic cycles in estrogen-treated mice. A radioautographic study. Expl Cell Res. *42:* 562–577 (1966).

Erickson, E.; Erickson, R.L.; Henry, B.; Pace, N.R.: Comparison of oligonucleotides produced by RNase T_1 digestion of 7,9 RNA from avian and purine oncorna-viruses and from unifected cells. Virology *53:* 40–46 (1973).

Errington, L.; Glass, R.E.; Hayward, R.S.; Scaife, J.G.: Structure and orientation of an RNA polymerase operon in *E. coli.* Nature, Lond. *249:* 519–522 (1974).

Evans, A.H.: Introduction of specific drug resistance properties by purified RNA-containing fractions from *Pneumococcus.* Proc. natn. Acad. Sci. USA *52:* 1442–1449 (1964).

Evans, M.I.; Hager, L.J.; McKnight, G.S.: A somatomedin-like peptide hormone is required during the estrogen-mediated induction of ovalbumin gene transcription. Cell *25:* 187–193 (1981).

Farber, E.: Biochemistry of carcinogenesis. Cancer Res. *28:* 2338–2349 (1968).

Fellenberg, G.; Schöemer, U.: Direct effect of IAA upon isolated chromatin of etiolated pea seedlings. Z. Pflanzenphysiol. *75:* 449–456 (1975).

Fend, M.M.; Villee, C.A.: Effect of RNA from estradiol-treated immature rats on protein synthesis in immature uteri. Endocrinology *88:* 279–285 (1971).

Ficq, A.; Brachet, J.: RNA-dependent DNA polymerase: possible role in the amplification of ribosomal DNA in *Xenopus* oocytes. Proc. natn. Acad. Sci. USA *68:* 2274–2318 (1971).

Flickinger, R.A.: Unbalanced growth and cell determination in frog embryos; in Niu, Segal, The role of RNA in reproduction and development, pp. 11–25 (North-Holland, Amsterdam 1973).

Fox, R.M.; Mendelson, J.; Barbosa, E.; Goulian, M.: RNA in nascent DNA from cultured human lymphocytes. Nature new Biol. *245:* 234–237 (1973).

Fraenkel-Conrat, H.; Singer, B.: Effect of small alkyl groups on m-RNA function. Proc. natn. Acad. Sci. USA *77:* 1983–1985 (1980).

Fraenkel-Conrat, H.; Singer, B.; Tsugita, A.: Purification of viral RNA by means of bentonite. Virology *14:* 54–58 (1961).

Franze-Fernandez, M.T.; Pogo, A.O.: Regulation of the nucleolar DNA-dependent RNA polymerase by amino-acids in Ehrlich ascitic tumor cells. Proc. natn. Acad. Sci. USA *68:* 3040–3044 (1971).

Frenster, J.H.: Correlation of the binding to DNA loops or to DNA helices with the effect on RNA synthesis. Nature, Lond. *206:* 1263–1270 (1965).

Frenster, J.H.: Selective control of DNA helix openings during gene regulation. Cancer res. *36:* 3394–3398 (1976).

Frenster, J.H.; Allfrey, V.G.; Mirsky, A.E.: Repressed and active chromatin isolated from interphase lymphocytes. Proc. natn. Acad. Sci. USA *50:* 1026–1032 (1963).

Fridland, A.; Brent, T.P.: DNA replication in methotrexate-treated human lymphoblasts. Eur. J. Biochem. *57:* 379–385 (1975).

Fridlender, B.; Fry, M.; Bolden, A.; Weissbach, A.: A new synthetic RNA-dependent DNA polymerase from human tissue culture. Proc. natn. Acad. Sci. USA *69:* 452–455 (1972).

Friend, C.; Freedman, H.A.E.: Effects and possible mechanism of action of dimethylsulfoxide on Friend cell differentiation. Biochem. Pharmac. *27:* 1309–1313 (1978).

Friend, C.; Scher, W.; Holland, J.G.; Sato, T.: Hemoglobin synthesis in murine virus induced leukemic cells in vitro: stimulation of erythroid differentiation by dimethylsulfoxide. Proc. natn. Acad. Sci. USA *68:* 378–382 (1971).

Froechner, S.C.; Bonner, J.: Ascites tumor ribonucleic acid polymerases. Isolation, purification and factor stimulation. Biochemistry *12:* 3064–3071 (1973).

Fujawara, Y.; Tatsumi, M.; Sasaki, M.: Cross-link repair in human cells and its possible defect in Fanconi's anaemia cells. J. molec. Biol. *113:* 635–649 (1977).

Fukui, Y.; Katsuma, H.: Dynamics of nuclear action bundle induction by dimethyl-sulfoxide and factors affecting its development. J. Cell Biol. *84:* 131–140 (1980).

Furth, J.E.; Natta, C.: Translational control of β and α globin chains synthesis. Nature new Biol. *240:* 274–276 (1972).

Gabbarro-Årpa, J.; Tougard, P.; Reiss, C.: Correlation of local stability of DNA during melting with environmental conditions. Nature, Lond. *280:* 515–517 (1979).

Gabbay, E.J.; Grier, D.; Fingerle, R.F.; Reimer, R.; Levy, R.; Pearce, S.W.; Wilson, W.D.: Interaction specificity of the anthracyclins with deoxyribonucleic acid. Biochemistry *15:* 2062–2070 (1976).

Gaffney, E.V.; Pigott, D.: Hydrocortisone stimulation of human mammary epithelial cells. In vitro *14:* 621–624 (1978).

Gal, A.; De Groot, N.; Hochberg, A.A.: The effect of dimethylsulfoxide on ribosomal fractions from rat liver. FEBS lett. *94:* 25–27 (1978).

Galand, P.; Dupont, N.: Biological activity of RNA from estrogen-stimulated uterus: in Niu, Segal, The role of RNA in reproduction and development, pp. 155–166 (North Holland, Amsterdam 1973).

Galand, P.; Leroy, F.; Chretien, J.: Effect of oestradiol and cell proliferation and histological changes in the uterus and vagina of mice. J. Endocr. *49:* 243–252 (1971).

Galand, P.; Rodesch, F.; Leroy, F.; Chretien, J.: Radio-autographic evaluation of the estrogen dependent proliferative pool in the stem cell compartment of the mouse uterine and vaginal epithelia. Expl Cell. Res. *48:* 595–604 (1967).

Gamper, H.B.; Straub, K.; Calvin, M.; Bartholomew, J.C.: DNA alkylation and unwinding induced by benzo(a)pyrene diol epoxide. Modulation by ionic strength and superhelicity. Proc. natn. Acad. Sci. USA *77:* 2000–2004 (1980).

Gannon, F.; Katzenellenbogen, B.; Stancel, G.; Gorski, J.: Estrogen-receptor movement to the nucleus. Discussion of a cytoplasmic exclusion hypothesis; in Papaconstantinou, The molecular biology of hormone action, pp. 137–149 (Academic Press, New York 1976).

Garnier, C.: Contribution à l'étude de la structure du fonctionnement des cellules glandulaires sereuses. J. Anat. Physiol., Paris, *47:* 22–98 (1900).

Garrigues, R.; Buu-Hoï, N.P.; Ramé, A.: Production de tumeurs végétales par action de la N-methyl-N-nitrosoaniline, composé cancérogène chez l'animal. C. r. hebd. Séanc. Acad. Sci., Paris, sér. D *273:* 1123–1125 (1971).

Gefter, M.L.: DNA polymerases II and III of *Escherichia coli.* Prog. nucleic Acid Res. mol. Biol. *14:* 101–115 (1974).

Gefter, M.L.: DNA replication. A. Rev. Biochem. *44:* 45–78 (1975).

Geider, K.; Berck, E.; Schaller, H.: An RNA transcribed from DNA at the origin of phage fd single strand to replicative form conversion. Proc. natn. Acad. Sci. USA *75:* 645–649 (1978).

Gelboin, H.V.; Klein, M.: Skin tumorogenesis by 7,12-dimethylbenz(a)anthracene: inhibition by actinomycin D. Science *145:* 1321–1322 (1964).

Gellert, M.; Mizuuchi, K.; O'Dea, M.H.; Nash, H.A.: DNA gyrase: an enzyme that produces superhelical turns into DNA. Proc. natn. Acad. Sci. USA *73:* 3872–3876 (1979).

Ghosh, N.K.; Cox, R.P.: Production of human chorionic gonadotropin in HeLa cell cultures. Nature, Lond. *259:* 416–417 (1976).

Ghosh, N.K.; Cox, R.P.: Induction of human follicle stimulating hormone in HeLa cells by sodium butyrate. Nature, Lond. *267:* 435–437 (1977).

Ghosh, N.K.; Rukenstein, A.; Cox, R.P.: Induction of human choriogonadotropin in HeLa cultures by aliphatic monocarboxylates and inhibitors of deoxyribonucleic acid synthesis. Biochem. J. *166:* 265–274 (1977).

Giacomoni, P.V.; Le Talaer, J.Y.; Le Peq, J.B.: *Escherichia coli* RNA polymerase binding sites on DNA are only 14 base pairs long and are located between sequences that are very rich in A+T. Proc. natn. Acad. Sci. USA *71:* 3091–3095 (1974).

Gierer, A.; Schramm, G.I.: Infectivity of ribonucleic acid tobacco mosaic virus. Nature, Lond. *177:* 702–703 (1956).

Gilbert, W.: Starting and stopping sequences for the RNA polymerase; in Losick, Chamberlin, RNA polymerase, pp. 193–205 (Cold Spring Harbor Press, New York 1976).

Gilbert, W.: Why genes in pieces? Nature, Lond. *271:* 501 (1978).

Gillum, A.H.; Clayton, D.A.: Mechanism of mitochondrial DNA-replication in mouse L-cells: RNA priming during the initiation of heavy strand synthesis. J. molec. Biol. *135:* 353–368 (1979).

Ginsburg, E.; Salmon, D.; Sreevalson, T.; Freese, E.: Growth inhibition and morphological changes caused by lipophilic acids in mammalian cells. Proc. natn. Acad. Sci. USA *70:* 2457–2461 (1973).

Githans, S.; Pictet, R.L.; Phelps, P.; Rutter, W.J.: 5-Bromodeoxyuridine may alter the differentiative program of the embryonic pancreas. J. Cell Biol. *71:* 341–356 (1976).

Glasser, S.R.; Chytil, F.; Spelsberg, T.C.: Early effects of oestradiol 17β on the chromatin and activity of the deoxyribonucleic acid-dependent ribonucleic acid polymerases (I and II) of the rat uterus. Biochem. J. *130:* 947–957 (1972).

Goldberg, I.H.; Friedman, P.A.: Antibiotics and nucleic acids. A. Rev. Biochem. *40:* 772–810 (1971).

Goldberg, I.H.; Reich, E.; Rabinowitz, M.: Inhibition of RNA polymerase reaction by actinomycin and proflavine. Nature, Lond. *199:* 44–46 (1963).

Goldberg, M.L.; Atchley, W.A.: The effect of hormones on DNA. Proc. natn. Acad. Sci. USA *55:* 989–996 (1966).

Goldstein, L.: Stable nuclear RNA returns to post-division nuclei following release to the cytoplasm during mitosis. Expl Cell Res. *89:* 421–425 (1974).

Goldstein, L.: Role of small nuclear RNAs in programming chromosomal information. Nature, Lond. *261:* 519–521 (1976).

Goldstein, E.I.; Penman, S.: Regulation of protein synthesis in mammalian cells. V. Effect of actinomycin D on translation control in HeLa cells. J. molec. Biol. *80:* 243–254 (1973).

Gorski, J.: Early estrogen effects on the activity of uterine ribonucleic acid polymerase. J. biol. Chem. *239:* 889–892 (1964).

Gorski, G.; Gannon, F.: Current models of steroid hormone action: a critic. A. Rev. Physiol. *38:* 425–450 (1976).

Gospodarowicz, D.: Localisation of a fibroblast growth factor and its effect alone and with hydrocortisone on 3T3 cell growth. Nature, Lond. *249:* 123–127 (1974).

Goulian, M.: Initiation of the replication of single stranded DNA by *Escherichia coli* DNA polymerase. Cold Spring Harb. Symp. quant Biol. *33:* 11–20 (1969).

Goulian, M.; Bleile, B.; Tseng, B.Y.: Methotrexate-induced misincorporation of uracil into DNA. Proc. natn. Acad. Sci. USA *77:* 1956–1960 (1980).

Grassé, P.P.: L'évolution du vivant, pp. 380–389 (Michel, Paris 1973).

Grassé, P.P.: Les genes en surimpression: une priorité. C. r. hebd. Séanc. Acad. Sci., Paris, sér. D *284:* 141–142 (1977).

Greenhouse, G.A.; Hynes, R.O.; Gross, P.R.: Sea urchin embryos are permeable to actinomycin. Science *171:* 686–689 (1971).

Greenleaf, A.L.; Krämer, A.; Bautz, E.K.F.: DNA-dependent RNA polymerases from *Drosophila melanogaster* larvae; in Losick, Chamberlin, RNA polymerase (Cold Spring Harbor Press, New York 1976).

Greenman, D.L.; Wicks, W.D.; Kenney, F.T.: Stimulation of ribonucleic acid synthesis by steroid hormone. II. High molecular weight components. J. biol. Chem. *240:* 4420–4426 (1965).

Gross, K.J.; Pogo, A.O.: Control of ribonucleic acid synthesis in eukaryotes. 2. The effect of protein synthesis on the activities of nuclear and total DNA-dependent RNA polymerases in yeast. Biochemistry *15:* 2070–2081 (1970).

Gross, P.R.; Cousineau, G.H.: Macromolecule synthesis and the influence of actinomycin on early development. Expl Cell Res. *33:* 368–395 (1969).

Gross, P.R.; Gross, K.W.: Gene transcription and gene expression during sea urchin development; in Niu, and Segal, The role of RNA in reproduction and development, pp. 4–10 (North Holland, Amsterdam 1973).

Groudin, M.; Weintraub, H.T.: Rous sarcoma virus activates embryonic globin genes in chicken fibroblasts. Proc. natn. Acad. Sci. USA *75:* 4464–4468 (1975).

Grouse, L.; Chilton, M.D.; McCarthy, B.J.: Hybridization of ribonucleic acid with unique sequence of mouse deoxyribonucleic acid. Biochemistry *11:* 798–805 (1972).

Grumm, F.G.; Armstrong, P.B.: Proteases are mitogenic to mesenchyme in vivo. Expl Cell Res. *119:* 317–326 (1979).

Grummt, J., Hall, S.H.; Crouch, R.J.: Localisation of an endonuclease specific for double stranded RNA within the nucleolus and its implication in processing ribosomal transcripts. Eur. J. Biochem. *94:* 437–443 (1979).

Grummt, J.; Smith, V.A.; Grummt, F.G.: Amino acid starvation affects the initiation frequency of nucleolar RNA polymerase. Cell *7:* 439–445 (1976).

Grunberger, D.; Pergolizzi, R.G.; Jones, R.E.: Translation of globin messenger RNA modified by benzo(a)pyrene 7,8-dihydrodiol-9,10-oxide in a wheat germ free system. J. biol. Chem. *255:* 390–394 (1980).

Guerriero, V., Jr.; Florini, J.R.: Dexamethasone effects on myoblast proliferation and differentiation. Endocrinology *106:* 1198–1202 (1980).

Guilfoyle, T.J.; Hanson, J.B.: Greater length of ribonucleic acid synthesized by chromatin-bound polymerase from auxin-treated soybean hypocotyls. Plant Physiol. *53:* 110–113 (1974).

Gurdon, J.B.: Adult frogs derived from the nuclei of single somatic cells. Devl Biol. *4:* 256–273 (1962).

Gusella, J.F.; Housman, D.: Induction of erythroid differentiation in vitro by purines and purine analogues. Cell *8:* 263–269 (1976).

Hadjan, A.J.; Kowarski, A.; Dickerman, H.W.; Migeon, C.J.: Specific tissue uptake and nuclear binding of testosterone by the erythropoietic mouse spleen. J. Steroid Biochem. *5:* 346–350 (1974).

Hagemann, R.F.: Effect of dimethylsulfoxide on RNA synthesis in S-180 tumor cells. Experientia *25:* 1298–1300 (1969).

Hager, G.; Holland, M.; Valenzuela, P.; Weinberg, F.; Rutter, W.J.: RNA polymerases and transcriptional specificity in *Saccharomyces cerevisiae;* in Losick, Chamberlin, RNA polymerase, pp. 745–762 (Cold Spring Harbor Press, New York 1976).

Haidle, C.W.: Fragmentation of deoxyribonucleic acid by bleomycin. Molec. Pharmacol. *7:* 645–652 (1971).
Haidle, C.W.; Bearden, J., Jr.: Effect of Bleomycin on an RNA-DNA hybrid. Biochem. biophys. Res. Commun. *65:* 815–821 (1965).
Hall, S.H.; Crouch, R.J.: Isolation and characterization of two enzymatic activities from chick embryos which degrade double-stranded RNA. J. biol. Chem. *252:* 4092–4097 (1977).
Hancock, R.L.; Forrester, P.I.; Lorscheider, F.L.; Lai, P.C.W.; Hay, D.M.: Ethionine-induced activation of embryonic genes; in Fishman, Selt, Oncodevelopmental gene expression. San Diego Conf. Oncodevelopment, pp. 247–251 (Academic Press, New York 1976).
Harris, G.: DNA synthesis and the production of antibodies by lymphoid tissues. Differentiation *1:* 301–317 (1973).
Harris, J.N.; Gorski, J.: Estrogen stimulation of DNA-dependent DNA polymerase activity in immature rat uterus. Mol. cell. Endocrinol. *10:* 293–305 (1978).
Harrison, P.R.: Analysis of erythropoiesis at the molecular level. Nature, Lond. *262:* 353–356 (1976).
Haskell, E.H.; Danern, C.I.: Pre-fork synthesis: a model for DNA replication. Proc. natn. Acad. Sci. USA *64:* 1065–1071 (1969).
Hawks, A.; Hicks, R.M.; Holsman, J.W.; Magel, P.N.: Morphological and biochemical effects of 1-2-dimethyl-hydrazine and 1-methyl-hydrazine in rats and mice. Br. J. Cancer *30:* 429–439 (1974).
Hecker, E.: Carcinogenic principles from the seed oil of Croton tiglium and from other *Euphorbiaceae.* Cancer Res. *28:* 2338–2349 (1968).
Heinrich, P.C.; Gross, V.; Northierman, W.; Scheurlen, M.: Structure and function of nuclear ribonucleoprotein complexes. Rev. Physiol. Biochem. Pharmacol. *81:* 101–134 (1978).
Helinski, D.R.: Plasmid DNA replication. Fed. Proc. *35:* 2026–2030 (1976).
Hellman, A.; Farrelly, J.G.; Martin, D.H.: Some biological properties of dimethylsulfoxide. Nature, Lond. *213:* 982–985 (1967).
Hennings, H.; Boutwell, R.K.: Studies on the mechanism of skin tumor promotion. Cancer Res. *30:* 312–320 (1970).
Henson, J.C.; Coune, A.; Heimann, R.: Cell proliferation induced by insulin in organ culture of the rat mammary carcinoma. Exl Cell Res. *45:* 351–360 (1966).
Herrick, G.; Albertz, B.: Nucleic acid helix-coil transitions mediated by helix-unwinding protein from calf thymus. J. biol. Chem. *251:* 2133–2141 (1976).
Herrick, G.; Delius, H.; Albertz, B.: Single-stranded DNA structure and DNA polymerase activity in the presence of nucleic acid helix unwinding proteins from calf thymus. J. biol. Chem. *251:* 2142–2146 (1976).
Herskowits, T.T.: Nonaqueous solutions of DNA: factors determining the stability of the helical configurations in solution. Archs Biochem. Biophys. *97:* 474–484 (1962).
Heywood, S.M.; Kennedy, D.S.; Bester, A.J.: Separation of specific initiation factors involved in the translation of myosin and myoglobin messenger RNAs and the isolation of a new RNA involved in translation. Proc. natn. Acad. Sci. USA *71:* 2428–2431 (1974).
Hilf, R.; Michel, I.; Silverstein, G.; Bell, C.: Effect of actinomycin D on estrogen-induced changes in enzymes and nucleic acids of R 3230 AC mammary tumors, uteri, and mammary glands. Cancer Res. *25:* 1854–1859 (1965).
Hillar, M.; Przyjenski, J.: Control of transcription and translation by low molecular weight peptides (deprimerones) from chromatin and poly(a) messenger RNA. Biochim. biophys. Acta *564:* 246–263 (1973).

Hinkle, D.; Chamberlin, M.J.: Studies on the binding of *E. coli* RNA polymerase to DNA. I. The role of sigma subunit. J. molec. Biol. *70:* 157–185 (1972).

Hiremath, S.T.; Wang, T.Y.: Effect of testosterone on RNA sequence complexity in rat prostate. Biochem. biophys. Res. Commun. *89:* 1200–1205 (1979).

Hirose, S.; Okazaki, R.; Tamanoi, F.: Mechanism of DNA chain growth. XI.Structure of RNA-linked DNA fragments of *Escherichia coli.* J. molec. Biol. *77:* 501–517 (1973).

Hobom, G.; Grosschedl, R.; Lusky, M.; Scherer, G.; Schwartz, E.; Kössel, H.: Functional analysis of the replicator structure of lambdoid bacteriophage DNA. Cold Spring Harb. Symp. quant. Biol. *43:* 165–178 (1978).

Holtfreter, J.: Neural differentiation of ectoderm through exposure to saline solution. J. exp. Zool. *95:* 307–340 (1944).

Holtzer, H.; Weintraub, H.; Biehl, J.: Cell cycle dependent events during myogenesis, neurogenesis and erythrogenesis; in Monroy, Tsanev, Biochemistry of cell differentiation, pp. 41–43 (Fed. Europ. Biochem. Soc. 7th Meet., Varna), (Academic Press, London 1973).

Honna, Y.; Kasakabe, T.; Hozumi, M.: Inhibition of differentiation of cultured mouse myeloid leukemia cells by non-steroid anti-inflammatory agents and counteraction of the inhibition by prostaglandin. Exp. Cancer Res. *39:* 2190–2194 (1979).

Horwitz, K.B.; McGuire, W.L.: Actinomycin D prevents nuclear processing of estrogen receptor. J. biol. Chem. *253:* 6319–6322 (1978).

Hough, B.R.; Davidson, E.H.: Studies on the repetitive sequence transcripts of *Xenopus* oocytes. J. molec. Biol. *70:* 491–509 (1972).

Howanessian, A.G.; Wood, J.; Meurs, E.; Montagniers, N.: Increased nuclease activity in cells treated with ppp A2′p5′A2′p5A. Proc. natn. Acad. Sci. USA *76:* 3261–3265 (1979).

Huberman, E.; Callaham, M.F.: Induction of terminal differentiation in human promyelocytic leukemia cells by tumor-promoting agents. Proc. natn. Acad. Sci. USA *76:* 1293–1297 (1979).

Huberman, J.A.; Riggs, A.D.: On the mechanism of DNA replication in mammalian chromosomes. J. molec. Biol. *32:* 327–341 (1968).

Hubinout, P.O.; Leroy, F.; Galand, P. (eds): Basic actions of sex steroids on target organs (Karger, Basel 1971).

Huggins, C.B.; Grand, L.: Neoplasms evoked in male Sprague-Dawley rats by pulse doses of 7,12-dimethylbenz(a)anthracene. Cancer Res. *26:* 2255–2258 (1966).

Humphries, S.; Windass, J.; Williamson, R.: Mouse globin gene expression in erythroid and non-erythroid tissues. Cell *7:* 267–277 (1976).

Hunt, T.; Ehrenfeld, E.: Cytoplasm from poliovirus-infected HeLa cells inhibits cell-free haemoglobin synthesis. Nature new Biol. *230:* 91–94 (1971).

Hyatt, E.A.: Polyriboadenylate synthesis by nuclei from developing sea urchin embryos. II. Polyriboadenylic acid priming of ATP polymerase. Biochim. biophys. Acta *142:* 254–262 (1967).

Iglewski, W.Y.; Franklin, R.M.: Purification and properties of reovirus ribonucleic acid. J. Virol. *1:* 302–307 (1967).

Ioannou, P.: General model for the replication of double stranded DNA molecules. Nature new Biol. *244:* 257–260 (1973).

Iriarte, P.V.; Hananian, J.; Corter, J.A.: Central nervous system leukemia and solid tumors of childhood. Treatment with 1,3-bis-(2-chloroethyl)-l-nitrosoureas (BCNU). Cancer, N.Y. *19:* 1187–1194 (1966).

Ishi, D.N.; Fibach, E.; Yamasaki, H.; Weinstein, B.I.: Tumor promoters inhibit morphological differentiation in cultured mouse neuroblastoma cells. Science *200:* 556–559 (1978).

Isono, K.; Yourno, J.: Chemical carcinogens as frameshift mutagens. *Salmonella* DNA sequence sensitive to mutagenesis by polycyclic carcinogens. Proc. natn. Acad. Sci. USA *71:* 1612–1617 (1974).

Iyer, V.N.; Szybalski, W.: A molecular mechanism of mitomycin action. Linking of complementary DNA strands. Proc. natn. Acad. Sci. USA *50:* 355–362 (1963).

Jacobson, A.; Firtel, R.A.; Lodish, H.F.: Transcription of polydeoxythymidylate sequences in the genome of the cellular slime mold: *Dictyostelium discoideum.* Proc. natn. Acad. Sci. USA *71:* 1607–1611 (1974).

Jahn, C.L.; Litman, G.W.: Accessibility of DNA in chromatin to the covalent binding of a chemical carcinogen; in Ahmad, Russell, Schultz, Werner, Miami Winter Symposia *15:* Differentiation and Development, p.492 (Academic Press, New York 1978).

Jensen, E.V.: Interaction of steroid hormones with the nucleus. Pharmac. Rev. *30:* 477–491 (1978).

Jensen, E.V.; Mohla, S.; Gorell, T.A.; Sombre, E.R. de: The role of estrophilin in estrogen action. Vitams Horm. *32:* 89–127 (1974).

Johnson, D.E.; McClure, W.R.: Abortive initiation on bacteriophage λ DNA; in Losick, Chamberlin, RNA polymerase, pp.413–418 (Cold Spring Harbor Press, New York 1975).

Johnson, L.F.; Lan, N.C.Y.; Baxter, J.D.: Evidence for a catalytic mechanism involved in the actions of glucocorticoid receptors on chromatin; in Ahmad, Russell, Schultz, Werner, Miami Winter Symposia *15:* Differentiation and Development, p.494 (Academic Press, New York 1978).

Johnston, F.P.; Jorgenson, K.F.; Lin, C.C.; Van de Sandle, J.H.: Interaction of anthracyclins with DNA and chromosomes. Chromosoma *68:* 115–129 (1978).

Jones, K.L.; Andersen, N.S.; Addison, J.: Glucocorticoid-induced growth inhibition of cells from a human lung alveolar cell carcinoma. Cancer Res. *38:* 1688–1693 (1978).

Kaempfer, R.: Identification and RNA-binding properties of an initiation factor capable of relieving translational inhibition induced by haem deprivation or double stranded RNA. Biochim. biophys. Res. Commun. *61:* 591–597 (1974).

Kaempfer, R.; Kaufman, J.: Inhibition of cellular protein synthesis by double-stranded RNA inactivation of an initiation factor. Proc. natn. Acad. Sci. USA *70:* 1222–1226 (1973).

Kalf, G.F.; Ch'ih, J.J.: Purification and properties of deoxyribonucleic acid polymerase from rat liver mitochondria. J. biol. Chem. *243:* 4904–4916 (1968).

Kandler-Singer, J.; Kathoff, K.: RNase sensitivity of an anterior morphogenetic determinant in an insect egg *(Smittia sp., Chirinomidae, Diptera).* Proc. natn. Acad. Sci. USA *73:* 3739–3743 (1976).

Kanehisa, T.; Oki, Y.; Ikuta, K.: Partial specificity of low molecular weight RNA that stimulates RNA synthesis in various tissues. Archs Biochem. Biophys. *165:* 146–152 (1974).

Kanehisa, T.; Tanaka, T.; Kano, Y.: Low molecular RNA associated with chromatin. Purification and characterization of RNA that stimulates RNA synthesis. Biochim. biophys. Acta *277:* 140–152 (1972).

Kang, C.Y.; Temin, H.M.: Endogenous RNA-directed DNA polymerase activity in uninfected chicken embryos. Proc. natn. Acad. Sci. USA *69:* 1550–1554 (1972).

Karlson, P.: New concepts on the mode of action of hormones. Perspect. Biol. Med. *6:* 203–214 (1963).

Kasamatsu, H.; Vinograd, J.: Replication of circular DNA in eukaryotic cells. A. Rev. Biochem. *43:* 695–719 (1974).

Kasukabe, T.; Honna, Y.; Hozumi, M.: The tumor promoter 12-O-tetradecanoyl-phorbol-13-

acetate inhibits or enhances induction of differentiation of mouse myeloid leukemia cells depending on the type of serum in the medium. Gann *72:* 310–314 (1981).

Kato, K.; Strauss, B.S.: Accumulation of an intermediate in DNA synthesis by Hep 2 cells treated with methylmethane sulfonate. Proc. natn. Acad. Sci. USA *71:* 1969–1973 (1974).

Katz, L.; Penman, S.: The solvent denaturation of double-stranded RNA from poliovirus infected HeLa cells. Biochem. biophys. Res. Commun. *23:* 557–560 (1966).

Kaufmann, Y.; Milcarek, C.; Berissi, H.; Penman, S.: HeLa cell poly(A)-mRNA codes for a subset of poly(A) + mRNA directed proteins with an actin as a major product. Proc. natn. Acad. Sci. USA *74:* 4801–4805 (1977).

Kawamata, J.; Nakabayashi, N.; Kawai, A.; Ushida, T.: Experimental production of sarcoma in mice with actinomycin. Med. J. Osaka Univ. *8:* 753–762 (1958).

Kay, E.R.M.: Incorporation of deoxyribonucleic acid by mammalian cells *in vitro.* Nature, Lond. *191:* 387–388 (1961).

Kedes, L.H.; Gross, P.R.; Cognetti, G.; Hunter, A.L.: Synthesis of nuclear and chromosomal proteins on light polyribosomes during cleavage in the sea urchin embryo. J. molec. Biol. *45:* 337–351 (1969).

Keir, H.M.; Craig, R.K.; McLennan, A.G.: Variation of deoxyribonucleic acid polymerases in the cell cycle. Symp. Biochem. Soc. *42:* 37–54 (1977).

Keller, W.: RNA primed DNA synthesis in vitro. Proc. natn. Acad. Sci. USA *69:* 1560–1564 (1972).

Keller, W.; Crouch, R.: Degradation of DNA-RNA hybrids by ribonuclease H and DNA polymerase of cellular and viral origin. Proc. natn. Acad. Sci. USA *69:* 3360–3364 (1972).

Kerr, I.M.; Brown, R.E.: pppA2′p5′A2′p5′A: An inhibitor of protein synthesis synthesized with an enzyme fraction from interferon-treated cells. Proc. natn. Acad. Sci. USA *75:* 256–260 (1978).

Kessler, B.: Interactions in vitro between gibberellins and DNA. II. Optical rotary profile of the thermal denaturation of DNA-gibberellin complexes. Biochim. biophys. Acta *232:* 611–613 (1971).

Kessler, B.; Snir, J.: Interactions in vitro between gibberellins and DNA. Biochim. biophys. Acta *195:* 207–218 (1969).

Khesin, R.B.; Astraurova, O.B.; Shermyakin, M.F.; Kanizolova, S.G.; Manyakov, V.F.: Changes in the properties of RNA polymerase upon binding to DNA and the initiation of RNA synthesis. Mol. Biol. *1:* 617–630 (1967).

Kimchi, A.; Shure, H.; Revel, M.: Antimitogenic function of interferon induced (2′5′)oligo(adenylate) and growth-related variations in enzymes that synthesize and degrade this oligonucleotide. Eur. J. Biochem. *114:* 5–10 (1981).

Kimchi, Y.; Palfrey, C.; Spector, I.; Barak, Y.; Littauer, U.Z.: Maturation of neuroblastoma cells in the presence of dimethylsulfoxide. Proc. natn. Acad. Sci. USA *73:* 462–466 (1976).

Kinsella, A.R.; Radman, M.: Tumor promoter induces sister chromatide exchanges; relevance to mechanisms of carcinogenesis. Proc. natn. Acad. Sci. USA *75:* 6149–6153 (1978).

Kleiman, L.; Huang, R.C.C.: Binding of actinomycin D to calf thymus chromatin. J. molec. Biol. *55:* 503–521 (1971).

Klevecz, R.; Hsu, T.C.: The differential capacity for RNA synthesis. A cytological approach. Proc. natn. Acad. Sci. USA *52:* 811–817 (1964).

Kluge, N.; Ostertag, W.; Sugizana, D.; Arndt-Jovin, D.; Steinheider, G.; Furusawa, M.; Dube, S.: Dimethylsulfoxide-induced differentiation and hemoglobin synthesis in tissue culture of rat erythroleukemia cells. Proc. natn. Acad. Sci. USA *73:* 1238–1240 (1976).

Kochakian, C.D.: Intracellular regulation of nucleic acids of mouse kidney by androgens. Gen. compar. Endocr. *13:* 146–150 (1969).

Kollar, E.J.; Baird, G.R.: Tissue interactions in embryonic mouse tooth germs. I. Reorganization of the dental epithelium. J. Embryol. exp. Morph. *24:* 159–171 (1970).

Kolodny, G.M.: Evidence for transfer of macromolecular RNA between mammalian cells in culture. Expl Cell Res. *65:* 313–324 (1971).

Kopelovich, L.: Are all normal diploid human cell strains alike? Relevance to carcinogenic mechanisms in vitro. Expl Cell Biol. *50:* 266–270 (1982).

Kopelovich, L.; Bias, N.E.; Helson, L.: Tumor promoter alone induces neoplastic transformation of fibroblasts from humans genetically predisposed to cancer. Nature, Lond. *282:* 619–620 (1979).

Kornberg, A.: DNA synthesis (Freeman, San Francisco 1974a).

Kornberg, R.D.: Chromatin structure. A repeating unit of histone and DNA. Science *184:* 868–871 (1974b).

Kottel, R.H.; Hoch, S.O.; Parsons, R.G.; Hoch, J.A.: Serum ribonuclease activity in cancer patients. Br. J. Cancer *38:* 280–286 (1978).

Krakow, J.S.; Ochoa, S.: Ribonucleic acid polymerase of *Azotobacter vinelandii.* I. Priming polyribonucleotides. Proc. natn. Acad. Sci. USA *49:* 88–94 (1963).

Krakow, J.S.; Rhodes, G.; Jovin, T.M.: RNA polymerase: catalytic mechanisms and inhibitors; in Losick, Chamberlin, RNA polymerase, pp. 127–157 (Cold Spring Harbor Press, New York 1976).

Krakow, J.S.; Van der Helm, K.: *Azotobacter vinelandii* RNA polymerase transition and the release of sigma. Cold Spring Harb. Symp. quant. Biol. *35:* 73–83 (1970).

Krey, A.K.; Allison, G.; Hahn, F.E.: Interaction of the antibiotic distamycin A with native DNA and with synthetic duplex polydeoxyribonucleotides. FEBS Lett. *29:* 58–62 (1973).

Krishman, I.; Baglioni, C.: Increased levels of (2′-5′)oligo(A)polymerase activity in human lymphoblastoid cells treated with glucocorticoids. Proc. natn. Acad. Sci. USA *77:* 6506–6510 (1980).

Kronenberg, L.H.; Humphreys, T.: Double stranded ribonucleic acid in sea urchin embryos. Biochemistry *11:* 2020–2026 (1972).

Krystosek, A.; Sachs, L.: Control of lysozyme induction in the differentiation of myeloid leukemic cells. Cell *9:* 675–684 (1976).

Lacassagne, A.: Apparition de cancer de la mammelle chez la souris mâle soumise à des injections de folliculine. C. r. hebd. Séanc. Acad. Sci., Paris *195:* 630–632 (1932).

Lacassagne, A.: Hormonal pathogenesis of adenocarcinoma of the breast. Am. J. Cancer *27:* 217–228 (1936).

Laemmli, V.K.: Levels of organization of the DNA in eukaryotic chromosomes. Pharmac. Rev. *30:* 469–476 (1979).

Lancelot, G.: Hydrogen bonding of amino acid side chains to nucleic acid bases. Biochimie *59:* 587–596 (1977a).

Lancelot, G.: Hydrogen bonding between nucleic acid bases and carboxylic acids. J. Am. chem. Soc. *99:* 7037–7042 (1977b).

Lancelot, G.; Helene, C.: Selective recognition of nucleic acids by proteins: the specificity of guanine interaction with carboxylate ions. Proc. natn. Acad. Sci. USA *74:* 4872–4875 (1977).

Lark, K.G.: Genetic control over the initiation of the synthesis of the short deoxyribonucleotide chains in *E. coli.* Nature new Biol. *240:* 237–240 (1972).

Larson, J.E.; Wells, R.D.: Neutropsin, a specific probe for A-T regions of duplex deoxyribonucleic acid. J. biol. Chem. *249:* 6719–6731 (1974).

Latt, S.A.; Wohlleb, J.C.: Optimal studies on the interaction of 33258 Hoechst with DNA, chromatin and metaphase chromosomes. Chromosoma *52:* 297–316 (1975).

Laval, J.: Two enzymes are required for strand scission in repair of alkylated DNA. Nature, Lond. *269:* 829–832 (1977).

Lavi, S.: Carcinogen-mediated amplification of viral DNA sequences in simian virus 40-transformed chinese hamster embryo cells. Proc. natn. Acad. Sci. USA *78:* 6144–6148 (1981).

Ledoux, L.; Charles, P.: Uptake of exogenous DNA by mouse embryos. Expl Cell Res. *45:* 498–501 (1967).

Lee, C.H.; Mizusawa, H.; Kakefuda, T.: Unwinding of double-stranded DNA helix by dehydration. Proc. natn. Acad. Sci. USA *78:* 2838–2842 (1981).

Lee, L.S.; Weinstein, I.B.: Mechanism of tumor promoter inhibition of cellular binding of epidermal growth factor. Proc. natn. Acad. Sci. USA *76:* 5168–5172 (1979).

Lee, S.C.; Dahmus, M.E.: Stimulation of eukaryotic DNA-dependent RNA-polymerase by protein factors. Proc. natn. Acad. Sci. USA *70:* 1383–1387 (1973).

Lee-Huang, S.; Sierra, J.M.; Naranjo, R.; Filipowitz, W.; Ochoa, S.: Eukaryotic oligonucleotides affecting m-RNA translation. Archs Biochem. Biophys. *180:* 276–287 (1977).

Le Goff, L.: Quelques propriétés d'*Agrobacterium tumefaciens* rendu résistent à la showdomycine. C. r. hebd. Séanc. Acad. Sci., Paris, sér. D *273:* 1757–1760 (1971).

Le Goff, L.; Aaron-da Cunha, M.I.; Beljanski, M.: Un ARN extrait d'*Agrobacterium tumefaciens* souches oncogenes et non oncogenes, élément indispensable à l'induction des tumeurs de *Datura stramonium.* Can. J. Microbiol. *22:* 694–701 (1976).

Le Goff, L.; Beljanski, M.: Stimulation de l'induction ou inhibition du développement des tumeurs de crown-gall par des ARN-fragments U_2. Interference de l'auxine. C.r. hebd. Séanc. Acad. Sci., Paris, sér. D *288:* 147–150 (1979).

Le Goff, L.; Beljanski, M.: In vivo stimulation and/or inhibition of Crown-gall tumors. Correlation with in vitro DNA synthesis, DNA strand separation and ribonuclease activity; in Lozano, Gwin, Cali, Proc. 5th Int. Conf. Plant Pathogenic Bact., pp. 295–307 (Columbia University of Missouri 1981).

Le Goff, L.; Beljanski, M.: Agonist and/or antagonist effects of plant hormones and an anticancer alkaloid on plant DNA structure and activity. IRCS med. Sci. *10:* 689–690 (1982).

Lehman, I.R.: DNA ligase: structure, mechanism and function. Science *186:* 790–797 (1974).

Lehman, I.R.; Bessman, M.J.; Simms, E.S.; Kornberg, A.: An enzymatic synthesis of deoxyribonucleic acid. I. Preparation of substrates and partial purification of an enzyme from *Escherichia coli.* J. biol. Chem. *233:* 163–170 (1958).

Leinwand, L.; Ruddle, F.H.: Stimulation of in vitro translation of messenger RNA by actinomycin D and cordycepin. Science *197:* 381–383 (1977).

Lentfer, D.; Lezins, A.G.: Mouse myeloma RNA polymerase B. Template specificities and the role of transcription stimulating factor. Eur. J. Biochem. *30:* 278–284 (1972).

Le Pecq, J.B.: Chimiotherapie anti-cancéreuse. Actualités scientifiques et industrielles *1388,* p. 39 (Hermann, Paris 1978).

Lerman, L.S.: Structural consideration in the interaction of DNA and acridines. J. molec. Biol. *3:* 18–30 (1961).

Leroy, F.; Bogaert, C.; Van Hoeck, J.: DNA replication and accelerated chain in oestrogen-stimulated uterine cells. Nature, Lond. *258:* 259–260 (1975).

Leshem, Y.: The molecular and hormonal basis of plant hormone regulation (Pergamon Press, Oxford 1973).

Leshem, Y.; Galston, A.W.: Reversion of isoperoxidase formation in excised tobacco pith by exogenous auxin-controlled RNA. Phytochemistry *10:* 2869–2878 (1971).

Levine, W.G.: Effect of dimethylsulfoxide on the hepatic disposition of chemical carcinogens. Ann. N.Y. Acad. Sci. *243:* 185–193 (1975).

Levine, L.; Gordon, J.A.; Jerncks, W.P.: The relationship of structure to the effectiveness of denaturing agents for deoxyribonucleic acid. Biochemistry *2:* 168–175 (1963).

Levy, J.; Terada, M.; Rifkind, R.A.; Marks, P.A.: Induction of erythroid differentiation by dimethylsulfoxide in cells infected with Friend virus. Relation to the cell cycle. Proc. natn. Acad. Sci. USA *72:* 28–32 (1975).

Lewis, B.; Abrell, J.; Smith, R.; Gallo, K.: Human DNA polymerase III (R-DNA polymerase): distribution from DNA polymerase and reverse transcriptase. Science *183:* 867–869 (1974).

Liao, S.: Cellular receptors and mechanism of action of steroid hormones. Int. Rev. Cytol. *41:* 87–172 (1975).

Liao, S.; Barton, R.W.; Lin, A.H.: Differential synthesis of ribonucleic acid in prostatic nuclei: evidence for selective gene transcription induced by androgens. Proc. natn. Acad. Sci. USA *55:* 1593–1600 (1966).

Liao, S.; Smythe, S.; Tomoczko, J.L.; Rossini, G.P.; Chen, C.; Hiipakka, R.A.: RNA-dependent release of androgen and other steroid-receptor complex from DNA. J. biol. Chem. *255:* 5545–5551 (1980).

Liao, S.; Tang, S.: Receptor proteins for androgen and the mode of action of androgens on gene transcription in ventral prostate. Vitams Horm. *27:* 17–90 (1969).

Lill, L.; Lill, U.; Sippel, A.; Hartmann, G.: The inhibition of the RNA-polymerase reaction by rifampicin; in Silvestri, RNA polymerase and transcription, pp. 55–64. Proc. 1st Int. Lepetit Colloq., Florence (North Holland, Amsterdam 1970).

Lilley, D.M.J.; Berendt, A.R.: The gross level of *in vitro* RNA synthesis in HeLa cells is unaltered by histone hyperacetylation. Biochem. biophys. Res. Commun. *90:* 917–924 (1979).

Lin, Y.C.; Loring, J.M.; Villee, C.A.: Diethylstilbestrol stimulates ornithine decarboxylase in kidney cells in culture. Biochem. biophys. Res. Commun. *35:* 1393–1403 (1980).

Lindell, T.J.: Evidence for an extranucleolar mechanism of actinomycin D action. Nature, Lond. *263:* 347–350 (1976).

Lindell, T.J.; O'Malley, A.F.; Puglisi, B.: Inhibition of nucleoplasmic transcription and the translocation of rapidly labeled nuclear proteins by low concentration of actinomycin D *in vivo.* Proposed role of messenger RNA in ribosomal transcription. Biochemistry *17:* 1154–1160 (1978).

Lindell, T.J.; Weinberg, F.; Morris, P.W.; Roeder, R.G.; Rutter, W.J.: Specific inhibition of nuclear RNA polymerase II by α-amanitin. Science *170:* 447–449 (1970).

Ling, V.: Pyrimidine sequences from the DNA of bacteriophage fd, fl and ΦX174. Proc. natn. Acad. Sci. USA *69:* 742–746 (1972).

Lippincott, J.A.; Lippincott, B.B.: Lysopine and octopine promote crown-gall tumor growth *in vivo.* Science *170:* 176–177 (1970).

Lippman, M.; Blan, G.; Huff, K.: The effects of estrogens and antiestrogens on hormone-responsive human breast cancer in long-term tissue culture. Cancer Res. *36:* 4595–4601 (1976).

Littau, V.C.; Allfrey, V.G.; Frenster, J.H.; Mirsky, A.L.: Active and inactive regions of nuclear chromatin as revealed by electron microscope autoradiography. Proc. natn. Acad. Sci. USA *52:* 93–100 (1964).

Liu, C.C.; Alberts, B.M.: Characterisation of RNA primer synthesis in the T_4 bacteriophage in vitro DNA replication system. J. biol. Chem. *256:* 2821–2829 (1981).

Liu, C.C.; Burke, R.L.; Hibner, U.; Barry, J.; Alberts, B.: Probing DNA replication mechanisms with the T4 bacteriophage *in vitro* system. Cold Spring Harb. Symp. quant. Biol. *43:* 469–487 (1978).

Lockwood, D.H.; Stockdall, F.E.; Topper, Y.J.: Hormone-dependent differentiation of mammary gland: sequence of action of hormones in relation to cell cycle. Science *156:* 945–946 (1967).

Loeb, L.A.; Agarwal, S.S.; Dube, D.K.; Gopinathan, E.P.; Shearman, G.S.C.; Sirover, M.A.; Travaglini, E.C.: Inhibition of mammalian DNA polymerases: possible chemotherapeutic approaches; in Sarin, Gallo, Inhibitors of DNA and RNA polymerases, pp. 27–46 (Pergamon Press, New York 1980).

Loeb, L.; Gelboin, H.V.: Stimulation of amino acid incorporation by nuclear ribonucleic acid from normal and methylcholanthrene-treated rats. Nature, Lond. *199:* 809–810 (1963).

Loeb, P.M.; Wilson, J.D.: Intranuclear localization of testosterone-1,2-H^3 in the preen gland of the duck. Clin. Res. *13:* 45 (1965).

Lohr, D.; Corden, J.; Tatchell, K.; Kovacio, R.T.; Holde, K.E. van: Comparative subunit structure of HeLa, yeast and chicken erythrocyte chromatin. Proc. natn. Acad. Sci. USA *74:* 79–83 (1977).

Losick, R.; Chamberlin, M.: RNA polymerase. Cold Spring Harbor Monogr. Ser. (New York 1976).

Lotem, J.; Sachs, L.: Regulation of normal differentiation in mouse and human myeloid leukemic cells by phorbol ester and the mechanism of tumor promotion. Proc. natn. Acad. Sci. USA *76:* 5158–5162 (1979).

Lowenhaupt, K.; Lingrel, J.V.: Synthesis and turnover of globin mRNA in murine erythroleukemia cells induced with haem. Proc. natn. Acad. Sci. USA *76:* 5173–5177 (1979).

Lowenstein, W.: The cell-to-cell membrane channel in development and growth; in Ahmad, Russell, Schultz, Werner, Miami Winter Symposia *15:* Differentiation and Development, pp. 399–401 (Academic Press, New York 1978).

Lown, J.W.; McLaughlin, L.W.; Chang, U.M.: Mechanism of action of haloethylnitrosoureas on DNA and its relation to their antileukemic properties. Bioorg. Chem. *7:* 97–110 (1978).

Luck, D.N.; Hamilton, T.H.: Early estrogen action. Stimulation of the metabolism of high molecular weight and ribosomal RNAs. Proc. natn. Acad. Sci. USA *69:* 157–161 (1972).

Ludlum, D.B.: Molecular biology of alkylation. An overview; in Sartorelli, Johns, Antineoplastic and immuno-suppressive agents, II, pp. 6–17 (Springer, Berlin 1975).

Lueck, J.D.; Nordlie, R.C.: Carbamyl phosphate: glucose-phosphotransferase activity of hepatic microsomal glucose-6-phosphatase. Biophys. biochem. Res. Commun. *39:* 190–196 (1970).

Magnan, F.R.; Neal, G.E.; Williams, D.C.: Subcellular distribution of testosterone in salt prostate and its possible relationship to nuclear ribonucleic acid synthesis. Archs Biochem. Biophys. *124:* 27–40 (1968).

Maitra, U.; Hurwitz, J.: The role of DNA in RNA synthesis. IX. Nucleoside triphosphate termini in RNA polymerase products. Proc. natn. Acad. Sci. USA *54:* 815–822 (1965).

Maitra, U.; Nakata, Y.; Hurwitz, J.: The role of deoxyribonucleic acid in ribonucleic acid synthesis. XIV. A study of the initiation of ribonucleic acid synthesis. J. biol. Chem. *242:* 4908–4918 (1967).

Marmur, J.; Anderson, W.F.; Matthews, L.; Berms, K.; Gajewska, E.; Doty, P.: The effects of ultraviolet light on the biological and physical chemical properties of deoxyribonucleic acids. J. cell. comp. Physiol. *58:* suppl. 1, pp. 33–55 (1961).

Marmur, J.; Grossman, L.: Ultraviolet light induced linking of deoxyribonucleic acid strands and its reversal by photoreactivating enzyme. Proc. natn. Acad. Sci. USA *47:* 778–787 (1961).

Martin, F.H.; Tinoco, I., Jr.: DNA-RNA hybrid duplexes containing oligo (dA:rU) sequences are exceptionally unstable and may facilitate termination of transcription. Nucl. Acids Res. *8:* 2295–2299 (1980).

Matsui, T.; Segall, J.; Weil, P.A.; Roeder, R.G.: Multiple factors for accurate initiation of transcription by purified RNA polymerase II. J. biol. Chem. *255:* 11992–11996 (1980).

Matsumoto, Y.; Yasuda, H.; Mita, S.; Marunoucha, T.; Yamada, M.: Evidence for the involvement of H_1 histone phosphorylation in chromosome condensation. Nature, Lond. *284:* 181–183 (1980).

Mayer, R.J.; Smith, R.G.; Gallo, R.C.: Reverse transcriptase in normal monkey placenta. Science *185:* 864–867 (1974).

Mayol, R.F.; Thayer, S.A.: Synthesis of estrogen-specific proteins in the uterus of the immature rat. Biochemistry *9:* 2484–2489 (1970).

Mazin, A.L.: Evolution of DNA structure. Direction, mechanism, rate. J. mol. Evol. *8:* 211–249 (1976).

McGuire, M.S.; Center, M.S.; Consigli, R.A.: Purification and properties of an endonuclease from nuclei of uninfected and polyoma-infected 3T3 cells. J. biol. Chem. *251:* 7746–7752 (1976).

McKnight, G.S.; Palmiter, R.D.: Transcriptional regulation of the ovalbumin and conalbumin genes by steroid hormones in chick oviduct. J. biol. Chem. *254:* 9050–9058 (1979).

McLeod, R.M.; King, C.E.; Hollander, V.P.: Effect of corticosteroids on ribonuclease and nucleic acid content of lymphosarcoma P 1798. Cancer Res. *23:* 1045–1058 (1963).

McPherson, A.; Jurnak, F.A.; Wang, A.H.J.; Molineux, I.; Rich, A.: Structure at 2.3 Å resolution of the gene 5 product of bacteriophage fd: a DNA unwinding protein. J. molec. Biol. *134:* 379–400 (1979).

Medda, A.K.; Dasmahapatra, A.K.; Ray, A.K.: Effect of estrogen and testosterone on the protein and nucleic acid content of liver, muscle and gonad and plasma protein content of male and female gonads (vitellogenic-nonvitellogenic) in *Singi* fish, Heteropneustes fossilis Bloch. Gen. compar. Endocr. *42:* 427–436 (1980).

Meehan, T.; Warshawsky, D.; Calvin, M.: Specific position involved in enzyme-catalysed covalent binding by benzo(a)pyrene to poly(G). Proc. natn. Acad. Sci. USA *73:* 1117–1122 (1976).

Meselson, M.; Stahl, F.V.: The replication of DNA. Cold Spring Harb. Symp. quant. Biol. *23:* 9–12 (1958).

Meyer, R.R.; Simpson, M.V.: Deoxyribonucleic acid biosynthesis in mitochondria. J. biol. Chem. *245:* 3426–3435 (1970).

Milanino, R.; Chargaff, E.: A purine polyribonucleotide synthetase from *Escherichia coli.* Proc. natn. Acad. Sci. USA *70:* 2558–2562 (1973).

Miller, L.; Brown, D.D.: Variation in the activity of nucleolar organizers and their ribosomal content. Chromosoma *28:* 430–444 (1969).

Mintz, B.; Illmensee, K.: Normal genetically mosaic mice produced from malignant teratocarcinoma cells. Proc. natn. Acad. Sci. USA *72:* 3585–3589 (1975).

Mirault, M.E.; Scherrer, K.: *In vitro* processing of HeLa cell preribosomes by nucleolar endonuclease. FEBS Lett. *20:* 233–238 (1972).

Mishra, N.C.; Niu, M.C.; Tatum, E.L.: Induction by RNA of inositol independence in *Neurospora crassa.* Proc. natn. Acad. Sci. USA *72:* 642–645 (1975).

Mohla, S.; De Sombre, E.R.; Jensen, E.V.: Tissue-specific stimulation of RNA synthesis by transformed estradiol-receptor complex. Biochem. biophys. Res. Commun. *46:* 661–667 (1972).

Mol, J.N.M.; Flavell, R.A.; Borst, P.: The presence of (dA-dT) 20-25 tracts in the DNA of primitive eukaryotes. Nucl. Acids Res. *3:* 2367–2377 (1976).

Molineux, I.J.; Friedman, S.; Gefter, M.L.: Purification and properties of the *Escherichia coli* deoxyribonucleic acid-unwinding protein. J. biol. Chem. *249:* 6090–6098 (1974).

Monod, J.: Le hasard et la nécessité (Le Seuil, Paris 1970).

Morgan, T.H.: Embryology and genetics (Columbia University Press, New York 1934).

Morris, C.N.R.; Morris, N.R.: A comparison of the structure of chicken erythrocyte and chicken liver chromatin. Cell *9:* 627–632 (1976).

Mukhopadhyay, D.K.; Dutta, S.K.: Evidence of increased amplification of ribosomal RNA genes in germinal conidial cells of *Neurospora crassa.* Indian J. exp. Biol. *17:* 620–621 (1979).

Müller, W.E.G.: Chromomycin, olivomycin and bleomycin as inhibitors of DNA and RNA polymerases. Int. Encyclop. Pharmacol. Ther. (1978).

Müller, W.E.G.; Yamazaki, Z.I.; Breter, H.J.; Zahn, R.K.: Action of Bleomycin on DNA and RNA. Eur. J. Biochem. *31:* 518–525 (1972).

Munck, A.; Foley, R.: Activation of steroid hormone-receptor complexes in intact target cell in physiological conditions. Nature, Lond. *278:* 752–754 (1979).

Myozis, R.K.; Grady, D.L.; Li, D.W.; Mirvis, S.E.; Ts'O', P.O.P.: Extensive homology of nuclear ribonucleic acid and polysomal (adenylic acid) messenger ribonucleic acid between normal and neoplastically transferred cells. Biochemistry *19:* 821–888 (1980).

Nagata, C.; Kodama, M.; Tagashira, Y.; Imamura, A.: Interaction of polynuclear aromatic hydrocarbons, 4-nitroquinoline-oxides and various dyes with DNA. Biopolymers *4:* 409–427 (1966).

Nagl, W.; Rücker, W.: Effects of phytohormones on thermal denaturation profiles of *Cymbidium* DNA: indication of differential replication. Nucl. Acids. Res. *3:* 2033–2039 (1976).

Nakata, Y.; Hurwitz, J.: The role of DNA in RNA synthesis. XIV. A study of the initiation of RNA synthesis. J. biol. Chem. *242:* 4908–4918 (1967).

Naveh-Many, T.; Cedar, H.: Actin gene sequences are undermethylated. Proc. natn. Acad. Sci. USA *78:* 4246–4250 (1981).

Neuhoff, V.; Schill, W.; Sternbach, H.: Microanalysis of pure deoxyribonucleic acid-dependent ribonucleic acid polymerase from *Escherichia coli.* Biochem. J. *117:* 623–631 (1970).

Neville, D.M., Jr.: Davies, D.R.: The interaction of acridine dyes with DNA: an X-ray diffraction and optical investigation. J. molec. Biol. *17:* 57–74 (1966).

Nilsen, T.; Baglioni, C.: Unusual base-pairing of newly synthesized DNA in HeLa cells. J. molec. Biol. *133:* 319–338 (1979).

Niu, M.C. in Rudnick, Cellular mechanisms in differentiation and growth, pp. 155–171 (Princeton University Press, Princeton 1956).

Niu, M.C.: The effect of mRNA on nuclear activity in developing systems; in Niu, Chuang, The role of RNA in development and reproduction, pp. 415–433 (Science Press, Beijing 1981).

Niu, M.C.; Niu, L.C.; Yang, S.F.: *In vivo* uptake of RNA and its function in the castrate uterus; in Niu, Segal, The role of RNA in reproduction and development, pp. 90–109 (North Holland, Amsterdam 1973).

Noll, M.; Kornberg, R.D.: Action of micrococcal nuclease on chromatin and the location of histone H_1. J. molec. Biol. *109:* 393–404 (1977).

Oberhauser, H.; Csordas, A.; Puschendorf, B.; Grunicke, H.: Increase in initiation sites for chromatin directed RNA synthesis by acetylation of chromosomal proteins. Biochem. biophys. Res. Commun. *84:* 110–116 (1978).

Okazaki, R.; Hirose, S.; Okazaki, T.; Ogawa, T.; Kurosawa, Y.: Assay of RNA-linked nascent DNA pieces with polynucleotide kinase. Biochem. biophys. Res. Commun. *62:* 1018–1024 (1975).

Okazaki, R.; Okazaki, T.; Sakabe, K.; Sugimoto, K.; Kainuma, R.; Sugino, A.; Iwatsuiki, N.: *In vivo* mechanism of DNA chain growth. Cold Spring Harb. Symp. quant. Biol. *33:* 129–143 (1968).

Okazaki, T.; Okazaki, R.: Mechanism of DNA chain growth. IV. Direction of synthesis of T4 short DNA chains as revealed by exonucleolytic degradation. Proc. natn. Acad. Sci. USA *64:* 1242–1248 (1969).

Olins, A.L.; Olins, D.E.: Spheroid chromatin units (U-bodies). Science *183:* 330–332 (1974).

O'Malley, B.W.; McGuire, W.L.: Progesterone induced synthesis of a new species of nuclear RNA. Endocrinology *84:* 63–68 (1969).

O'Malley, B.; Schwartz, R.J.; Schrader, W.T.: A review of regulation of gene expression by steroid hormone receptors. J. Steroid Biochem. *7:* 1151–1159 (1976).

Ono, T.; Culter, R.G.: Age-dependent relaxation of gene repression. Increase of endogenous murine-leukemia virus-related and globin-related RNA in brain and liver of mice. Proc. natn. Acad. Sci. USA *75:* 4431–4435 (1978).

Otto, A.M.; Zumbe, A.; Gibson, L.; Kubler, A.M.; De Asua, L.J.: Cytoskeleton-disrupting drugs enhance effect of growth factors and hormones on initiation of DNA synthesis. Proc. natn. Acad. Sci. USA *76:* 6435–6438 (1979).

Paffenholz, V.; Lee, H.V.; Ho, Y.K.; Bardos, T.J.: Uptake of partially thiolated DNA by ascites tumor cells. Cancer Res. *36:* 1445–1452 (1976).

Pagano, J.S.: Biological activity of isolated viral nucleic acids; in Melnick, Prog. med. Virol., vol. 12, pp. 1–48 (Karger, Basel 1970).

Palmiter, R.D.; Schimke, R.T.: Regulation of protein synthesis in chick oviduct. J. biol. Chem. *248:* 1502–1512 (1973).

Patel, G.L.; Thompson, P.E.: Immunoreactive helix-destabilizing protein localized in transcriptionally active regions of *Drosophila* polytene chromosomes. Proc. natn. Acad. Sci. USA *77:* 6749–6753 (1980).

Paterson, B.M.; Bishop, J.O.: Changes in the mRNA population of chick myoblasts during myogenesis *in vitro.* Cell *12:* 751–765 (1977).

Pauley, R.J.; Socher, S.H.: Hormonal influences on the expression of casein messenger RNA during mouse mammary tumorigenesis. Cancer Res. *40:* 362–367 (1980).

Pauley, R.J.; Sosen, J.M.; Socher, S.H.: Mammary tumor virus and casein gene transcription during mouse mammary development. Nature, Lond. *275:* 455–457 (1978).

Peackocke, A.R.; Skerrett, J.N.H.: The interaction of amino acridines with nucleic acids. Trans. Faraday Soc. *52:* 261–279 (1976).

Pelling, G.: Chromosomal synthesis of ribonucleic acid as shown by incorporation of uridine labelled with tritium. Nature, Lond. *184:* 655–656 (1959).

Penon, P.; Tiessere, M.; Azou, Y.; Ricard, J.: Controle hormonal de la transcription nucléolaire chez les végétaux supérieurs. Physiol. vég. *1:* 813–829 (1975).

Perdue, J.F.; Lubenskyi, W.; Kivity, E.; Sonder, S.A.; Fenton, J.W.: Protease mitogenic response of chick embryo fibroblasts and receptor binding processing of human α-thrombin. J. biol. Chem. *256:* 2767–2776 (1981).

Perlman, S.W.; Ford, P.J.; Rosbach, M.M.: Presence of tadpole and adult globin sequences in oocytes of *Xenopus laevis.* Proc. natn. Acad. Sci. USA *74:* 3835–3839 (1977).

Perry, R.P.: Processing of RNA. A. Rev. Biochem. *45:* 605–629 (1976).

Perry, R.P.; Kelley, D.E.; Schibler, V.; Huebner, K.; Croce, C.M.: Selective suppression of the transcription of ribosomal genes in mouse-human hybrid cells. J. cell. Physiol. *98:* 553–560 (1979).

Petricciani, J.C.; Patterson, R.M.: Incorporation of exogenous DNA into mammalian chromosomes. Nature, Lond. *249:* 649–670 (1974).

Philipson, L.; Wall, R.; Glickman, R.; Darnell, J.: Addition of polyadenylate sequences to virus-specific RNA during adenovirus replication. Proc. natn. Acad. Sci. USA *68:* 2806–2809 (1971).

Pieczenic, G.; Horiouchi, K.; Model, P.; McGill, C.; Mazur, B.J.; Vovis, G.F.; Zinder, N.D.: Is mRNA transcribed from the strand complementary to it in a DNA duplex? Nature, Lond. *253:* 131–132 (1975).

Pigiet, V.; Eliasson, R.; Reichard, P.: Replication of polyoma DNA in isolated nuclei. III. The nucleotide sequence at the RNA-DNA junction of nascent strands. J. molec. Biol. *84:* 197–216 (1974).

Plawecki, M.; Beljanski, M.: Transcription par la polynucleotide phosphorylase de l'ARN associé à l'ADN d'*Escherichia coli.* C.r. hebd. Séanc. Acad. Sci., Paris, sér. D *273:* 827–830 (1971).

Plawecki, M.; Beljanski, M.: Synthèse *in vitro* d'un ARN comme amorceur pour la réplication de l'ADN. C.r. hebd. Séanc. Acad, Sci., Paris, sér. D *278:* 1413–1416 (1974).

Plawecki, M.; Beljanski, M.: Comparative study of *Escherichia coli* endotoxin, hydrocortisone and Beljanski Leukocyte Restorer activity in cyclophosphamide-treated rabbits. Proc. Soc. exp. Biol. Med. *168:* 408–413 (1981).

Plumbridge, T.W.; Brown, J.R.: Studies on the mode of interaction of 4′-epi-adriamycin and 4-demethoxy-daunorubicin with DNA. Biochem. Pharmac. *27:* 1881–1882 (1978).

Pollard, T.D.; Weihing, R.R.: Actin and myosin and cell movement. CRC crit. Rev. Biochem. *2:* 1–65 (1974).

Pontecorvo, G.: Microbial genetics: retrospect and prospect. Proc. R. Soc., biol. Sci., Lond., ser. B *158:* 1–23 (1963).

Pottathil, R.; Meier, H.: Antitumor effects of RNA isolated from murine tumors and embryos. Cancer Res. *37:* 3280–3286 (1977).

Prakash, L.; Strauss, B.: Repair of alkylation damage: stability of methyl groups in *Bacillus subtilis* treated with methyl membrane sulfonate. J. Bact. *102:* 760–766 (1970).

Ptashne, M.; Backman, K.; Humayun, M.Z.; Jeffrey, A.; Maurer, R.; Meyer, B.; Sauer, R.T.: Autoregulation and function of a repressor in bacteriophage lambda. Interactions of a regulatory protein with sequences in DNA mediate intricate patterns of gene regulation. Science *194:* 156–161 (1976).

Quigley, J.P.: Phorbol ester-induced morphological changes in transformed chick fibroblasts: evidence for direct catalytic involvement of plasminogen activation. Cell *17:* 131–141 (1979).

Raff, R.A.; Colot, H.V.; Selvig, S.E.; Gross, P.R.: Oogenetic origin of messenger RNA for embryonic synthesis of microtubule proteins. Nature, Lond. *235:* 211–214 (1972).

Raff, R.A.; Greenhouse, G.; Gross, K.W.; Gross, P.R.: Synthesis and storage of microtubule proteins by sea urchin embryos. J. Cell Biol. *50:* 516–527 (1971).

Raick, A.N.: Ultrastructural, histological and biochemical alteration produced by 12-O-tetradecanoyl-phorbol-13-acetate on mouse epidermis and their relevance to skin tumor promotion. Cancer Res. *33:* 269–286 (1973).

Rammler, D.H.; Zaffaroni, A.: Biological implication of DMSO based on review of its chemical properties. Ann. N.Y. Acad. Sci. *141:* 13–23 (1967).

Raynaud-Jammet, C.; Catelli, M.G.; Baulieu, E.E.: Inhibition by α-amanitin of the oestradiol-induced increase in α-amanitin insensitive RNA polymerase in immature rat uterus. FEBS Lett. *22:* 93–96 (1972).

Reddi, K.K.: Ribonuclease induction in cells transformed by *Agrobacterium tumefaciens.* Proc. natn. Acad. Sci. USA *56:* 1207–1214 (1966).

Reich, E.: Actinomycin: correlation of structure and function of its complexes with purines and DNA. Science *143:* 684–689 (1964).

Reich, E.; Franklin, R.M.; Shatkin, A.J.; Tatum, E.L.: Effect of actinomycin D on cellular nucleic acid synthesis and virus production. Science *134:* 556–557 (1961).

Reich, E.; Goldberg, H.: Actinomycin and nucleic acid function. Prog. nucleic Acid Res. mol. Biol. *3:* 183–234 (1970).

Reichman, M.; Penman, S.: Stimulation of polypeptide initiation *in vitro* after protein synthesis inhibition *in vivo* in HeLa cells. Proc. natn. Acad. Sci. USA *70:* 2678–2682 (1973).

Rein, A.: The small molecular weight monodisperse nuclear RNAs in mitotic cells. Biochim. biophys. Acta *232:* 306–313 (1971).

Rein, D.; Gruenstein, E.; Lessard, J.: Actin and myosin synthesis during differentiation of neuroblastoma cells. J. Neurol. Neurosurg. Psychiat. *34:* 1459–1469 (1980).

Resch, K.; Bouillon, D.; Gemsa, D.; Averdunk, R.: Drugs which disrupt microtubules do not inhibit the initiation of lymphocyte activation. Nature, Lond. *265:* 349–351 (1977).

Rhodes, G.; Chamberlin, M.J.: A kinetic analysis of RNA chain initiation by *Escherichia coli* ribonucleic acid polymerase bound to DNA. J. biol. Chem. *250:* 9112–9120 (1975).

Richardson, J.P.: The binding of RNA polymerase to DNA. J. molec. Biol. *21:* 83–114 (1966).

Richardson, J.P.: RNA polymerase and the control of RNA synthesis. Prog. nucleic Acid Res. mol. Biol. *9:* 75–116 (1969).

Richeman, R.A.; Claus, T.H.; Pilkis, S.J.; Friedman, D.L.: Hormonal stimulation of DNA synthesis in primary culture of adult rat hepatocytes. Proc. natn. Acad. Sci. USA *73:* 3589–3591 (1976).

Ris, H.: Chromosomal structure as seen by electron microscopy. Ciba Fdn Symp. *28:* 7–15 (1975).

Roeder, R.G.: Eukaryotic nuclear RNA polymerases. A. Rev. Biochem. *44:* 285–329 (1975).

Roeder, R.G.; Rutter, W.J.: Specific nucleolar and nucleoplasmic RNA polymerases. Proc. natn. Acad. Sci. USA *65:* 675–682 (1970).

Romanof, A.L.: The avian embryo (MacMillan, New York 1960).

Rosen, J.M.; Socher, S.H.: Detection of casein messenger RNA in hormone dependent mammary cancer by molecular hybridization. Nature, Lond. *269:* 83–86 (1977).

Roussaux, J.: Activité néoformatrice d'acides ribonucléiques extraits de tumeurs végétales. Physiol. Plant. *35:* 269–272 (1975).

Rousseau, G.G.: Interaction of steroids with hepatoma cells: molecular mechanisms of glucocorticoid hormone action. J. Steroid Biochem. *6:* 75–89 (1975).

Rousseau, G.G.; Baxter, J.D.; Aiggins, S.J.; Tomkins, G.M.: Steroid-induced nuclear binding of glucocorticoid receptors in intact hepatoma cells. J. molec. Biol. *79:* 539–554 (1973).

Roven, L.; Kornberg, A.: Primase, the dnaG protein of *Escherichia coli.* An enzyme which starts DNA chains. J. biol. Chem. *253:* 770–774 (1978).

Rovera, G.; O'Brien, T.; Diamond, L.: Tumor promoters inhibit spontaneous differentiation of Friend erythroleukemia cells in culture. Proc. natn. Acad. Sci. USA *74:* 2894–2896 (1977).

Ruiz-Carillo, A.; Wangh, L.J.; Allfrey, V.G.: Processing of newly synthesized histone molecules. Nascent histone H_4 chains are reversibly phosphorylated and acetylated. Science *190:* 117–128 (1975).

Rusch, H.P.: Some biochemical events in the life cycle of *Physarum polycephalum;* in Prescott, Golstein, McConkey, Advances in cell biology, vol. 1, pp. 297–327 (Appleton-Century-Crofts, New York 1970).

Rutter, W.J.; McDonald, R.J.; Van Nest, G.; Harding, J.D.; Przybyla, A.E.; Chirgwin, J.M.; Pictet, R.L.: Pancreas specific genes and their expression during differentiation; in Ahmad, Russell, Schultz, Werner, Miami Winter Symposia *15:* Differentiation and Development, pp. 65–91 (Academic Press, New York 1978).

Ryser, H.J.P.: Chemical carcinogenesis. New Engl. J. Med. *285:* 721–734 (1971).

Sagher, D.; Harvey, R.G.; Hsu, W.T.; Weiss, S.B.: Effect of benzo(a)pyrene-diolepoxide on infectivity and in vitro translation of phage MS2 RNA. Proc. natn. Acad. Sci. USA *76:* 620–624 (1979).

Samuels, H.H.; Shapiro, L.E.: Thyroid hormone stimulates *de novo* growth hormone synthesis in cultured GH_1 cells: evidence for the accumulation of a rate limiting species in the induction process. Proc. natn. Acad. Sci. USA *73:* 3369–3373 (1976).

Sandberg, A.A.; Sakurai, M.: Chromosomes in the causation and progression of cancer and leukemia; in Busch, The molecular biology of cancer, pp. 81–106 (Academic Press, New York 1974).

Sarasin, A.; Moulé, Y.: Helical polysomes induced by aflatoxin B_1 *in vivo.* Expl Cell Res. *97:* 346–358 (1976).

Sarkander, H.I.; Fleischer-Lambropoulis, H.; Bradé, W.P.: A comparative study of histone acetylation in neural and glial nuclei enriched rat brain fractions. FEBS Lett. *52:* 40–43 (1973).

Sarkander, H.I.; Knoll-Köhler, E.: Changing patterns of histone acetylation and RNA synthesis of the developing and ageing rat brain. FEBS Lett. *85:* 301–304 (1978).

Sarma, D.S.R.; Rajalakshimi, M.; Farber, E.: Chemical carcinogenesis. Interactions of carcinogens with nucleic acids; in Becker, Cancer, a comprehensive treatise, vol. 1, pp. 235–271 (Plenum Press, New York 1975).

Sartiano, G.P.; Lynch, W.E.; Bullington, W.D.: Mechanism of action of the anthracycline antitumor antibiotics, docorubicin, daunomycin and rubidazone: preferential inhibition of DNA polymerase α. J. Antibiot., Tokyo, *32:* 1038–1045 (1979).

Sasaki, N.; Niu, M.C.: The role of RNA in the differentiation of presumptive ectoderm from urodele embryos; in Niu, Segal, The role of RNA in reproduction and development, pp. 183–198 (North Holland/American Elsevier, New York 1973).

Saxén, L.; Karkiner-Jaeskelainer, M.; Lehtonen, E.; Nordling, S.; Wartiovaara, J.: in Poste, Nicolson, The cell surface in animal embryogenesis and development, pp. 331–408 (North Holland/American Elsevier, New York 1976).

Schadler, W.T.; Toft, D.O.; O'Malley, B.W.: Progesterone-binding of chick oviduct. J. biol. Chem. *247:* 2401–2407 (1972).

Schaller, H.; Gray, C.; Hermann, K.: Nucleotide sequence of an RNA polymerase binding site from the DNA of bacteriophage fd. Proc. natn. Acad. Sci. USA *72:* 737–741 (1975).

Schekman, R.; Wickner, W.; Westergaard, O.; Brutlag, D.; Geider, K.; Bertsch, L.L.; Kornberg, A.: Initiation of DNA synthesis of ΦX174 replication form requires RNA synthesis resistant to rifampicin. Proc. natn. Acad. Sci. USA *69:* 2691–2695 (1972).

Schimamito, N.; Wu, C.W.: Mechanism of ribonucleic acid chain initiation. 1. A non-steady state study of ribonucleic acid synthesis without enzyme turnover. Biochemistry *19:* 842–848 (1980).

Schimke, R.T.: Protein synthesis and degradation in animal tissue; in Paul, Biochemistry of cell differentiation, vol. 9, pp. 183–221 (University Park Press, Baltimore 1974).

Schlager, S.I.; Paque, R.E.; Dray, S.: Complete and apparently specific local tumor regression using syngeneic or xenogeneic 'tumor-immune' RNA extracts. Cancer Res. *35:* 1907–1914 (1975).

Schmidt, A.; Zilberstein, A.; Shulman, L.; Federman, P.; Berissi, H.; Revel, M.: Interferon action: isolation of nuclease F, a translation inhibition activated by interferon-induced (2′-5′)oligo-isoadenylate. FEBS Lett. *95:* 257–264 (1978).

Schodell, R.: Effect of different media supplements on amino acid uptake and mitotic stimulation in BHK cells. Nature new Biol. *243:* 83–85 (1973).

Schrecker, A.W.; Smith, R.G.; Gallo, R.C.: Comparative inhibition of purified DNA polymerases from murine leukemia virus and human lymphocytes by 1-β-arabinofumaro-sylcytosine-5′-triphosphate. Cancer Res. *34:* 286–292 (1974).

Schwartz, H.S.: Some determinants of the therapeutic efficacy of actinomycin D, adriamycin and daunorubicin. Cancer Chemother. Rep. *58:* 55–62 (1974).

Scrutton, M.C.; Wu, C.W.; Goldthwait, D.A.: The presence and possible role of zinc in RNA polymerase obtained from *Escherichia coli.* Proc. natn. Acad. Sci. USA *68:* 2497–2501 (1971).

Seeds, N.W.; Gilman, A.G.; Amano, T.; Nirenberg, M.W.: Regulation of axon formation by clonal lines of a neural tumor. Proc. natn. Acad. Sci. USA *66:* 160–167 (1970).

Segal, S.J.; Ige, R.; Burgos, M.; Tuohimao, P.; Koisz, S.S.: Specific and heterospecific transfer of hormone action by mRNA; in Niu, Segal, The role of RNA in reproduction and development, pp. 270–283 (North Holland, Amsterdam 1973).

Seifart, K.H.; Benecke, B.J.; Jahasz, P.P.: Multiple RNA polymerase species from rat liver tissue. Possible existence of a cytoplasmic enzyme. Archs Biochem. Biophys. *151:* 519–532 (1972).

Seifart, K.H.; Juhasz, P.P.; Benecke, B.J.: A protein factor from rat liver tissue enhancing the transcription of native templates by homologous RNA polymerase B. Eur. J. Biochem. *33:* 181–191 (1973).

Sensenbrenner, M.; Jaros, G.G.; Moonen, G.; Mandel, P.: Effects of synthetic tripeptide on the differentiation of dissociated cerebral hemisphere nerve cells in culture. Neurobiology. *5:* 207–213 (1975).

Sharma, O.K.; Borek, E.: The carcinogen ethionine elevates progesterone levels. Nature, Lond. *265:* 748–749 (1977).

Shatkin, A.J.: Capping of eucaryotic mRNAs. Cell *9:* 645–653 (1976).

Shen, S.C.; Hong, M.; Cai, R.; Chen, W.; Chang, W.: Drug resistance induced in *B. subtilis* by RNA Sci. sinica *11:* 233 (1962).

Shenkin, A.; Burdon, R.H.: Deoxyadenylate-rich and deoxyguanylate-rich regions in mammalian DNA. J. molec. Biol. *85:* 19–39 (1974).

Shih, T.Y.; Bonner, J.: Chromosomal RNA of calf thymus chromatin. Biochim. biophys. Acta *182:* 30–35 (1969).

Shimizu, N.; Sokawa, Y.: 2′,5′-Oligoadenylate synthetase activity in lymphocytes from normal mouse. J. biol. Chem. *254:* 12034–12037 (1979).

Show, B.; Herman, T.; Kovacic, R.; Beaudreau, G.; Holde, K. van: Analysis of subunit organization in chicken erythrocyte chromatin. Proc. natn. Acad. Sci. USA *73:* 505–509 (1976).

Shoyab, M.: Dose-dependent preferential binding of polycyclic aromatic hydrocarbons to reiterated DNA of murine skin cells in culture. Proc. natn. Acad. Sci. USA *75:* 5841–5845 (1978).

Shugalin, A.V.; Frank-Kamanetstii, A.D.; Lazarkin, Y.S.: Viscometric investigation of block heterogeneity of DNA. Mol. Biol. *4:* 221–227 (1970).

Shyamala, G.; Gorski, J.: Estrogen receptors in the rat uterus. Studies on the interaction of cytosol and nuclear binding sites. J. biol. Chem. *244:* 1097–1103 (1969).

Shyamala-Harris, G.: Nature of estrogen specific binding sites in the nuclei of mouse uteri. Nature, Lond. *231:* 246–248 (1971).

Siddiqui, M.A.Q.; Arnold, H.H.; Deshpande, A.K.; Jakowlew, S.B.; Crawford, P.A.: Embryonic gene regulation: role of an inducer RNA in manipulation of embryonic gene functions. Archs. Biol. Méd. exper. *12:* 331–348 (1979).

Sieber, F.; Stuart, R.K.; Spivak, J.L.: Tumor-promoting phorbol esters stimulate myelopoiesis and suppress erythropoiesis in cultures of mouse bone marrow cells. Proc. natn. Acad. Sci. USA *78:* 4402–4406 (1981).

Sigal, N.; Delius, H.; Kornberg, T.; Gefter, M.; Alberts, B.: A DNA unwinding protein isolated from *Escherichia coli.* Its interaction with DNA and DNA polymerases. Proc. natn. Acad. Sci. USA *69:* 3537–3541 (1972).

Silberberg, M.; Silberberg, R.: Leukomogenic action of adenocorticotropic hormone (ACTH) in mice of variant age. Cancer Res. *15:* 291–293 (1955).

Singer, B.: Sites in nucleic acids reacting with alkylating agents of differing carcinogenicity or mutagenicity. J. Toxicol. envir. Health *2:* 1279–1295 (1977).

Sippel, A.E.; Hartmann, G.: Rifampicin resistance of RNA polymerase in the binary complex with DNA. Eur. J. Biochem. *16:* 152–157 (1970).

Sivak, A.; Van Duuren, B.L.: RNA synthesis induction in cell culture by a tumor promoter. Cancer Res. *30:* 1203–1205 (1970).

Sklar, V.E.F.; Roeder, R.G.: Purification, characterization and structure of class III RNA polymerases. Fed. Proc. *34:* 650 (1975).

Slaga, T.Y.; Scribner, J.D.; Thompson, S.; Viaje, A.: Epidermal cell proliferation and promoting ability of phorbol esters. J. natn. Cancer Inst. *57:* 1145–1149 (1976).

Slavkin, H.C.; Bringas, P.; Bavetta, L.A.: Ribonucleic acid within the extracellular matrix during embryonic tooth formation. J. cell. Physiol. *73:* 179–190 (1969).

Slavkin, H.C.; Croissant, R.: Intercellular communication during odontogenic epithelial-mesenchymal interactions: isolation of extracellular matrix vesicles containing RNA; in Niu, Segal, The role of RNA in reproduction and development, pp. 247–258 (North Holland/American Elsevier, New York 1973).

Slavkin, H.C.; Flores, P.; Bringas, P.; Bavetta, L.A.: Epithelial-mesenchymal interactions during odontogenesis. I. Isolation of several intercellular matrix low molecular weight methylated RNAs. Devl Biol. *23:* 276–296 (1971).

Slavkin, H.C.; Zeichmer-David, M.: The possible mode of transmission for 'inductive RNA' during epithelial-mesenchymal interactions; in Niu, Chuang, The role of RNA in develop-

ment and reproduction (Science Press, Beijing 1981; Van Nostrand & Reinhold, London 1981).
Slifkin, M.; Merkow, L.P.; Pardo, M.; Epstein, S.M.; Leighton, J.; Farber, E.: Growth *in vitro* of cells from hyperplasic nodules of liver induced by 2-fluorenylacetamide or aflatoxin B_1. Science *167:* 285–287 (1970).
Smith, C.G.: Differential interaction of nogalamycin with DNA of varying base composition. Proc. natn. Acad. Sci. USA *54:* 566–572 (1965).
Smulson, M.E.; Sudhakar, S.; Tew, K.D.; Butt, T.R.; Jump, D.B.: The influence of nitrosoureas on chromatin nucleosomal structure and function; in Busch, Crooke, Daskal, Effects of drugs on the cell nucleus, pp. 333–357 (Academic Press, New York 1979).
Sobell, H.M.: The stereochemistry of actinomycin binding to DNA and its implication in molecular biology; in Davidson, Cohn, Progress in nucleic acid research and molecular Biology, vol. 13, pp. 153–190 (Academic Press, New York 1973).
Sobell, H.M.; Jain, S.C.: Stereochemistry of actinomycin binding to DNA. II. Detailed molecular model of actinomycin-DNA complex and its implications. J. molec. Biol. *68:* 21–34 (1972).
Sobota, A.E.: Effect of sublethal heat injury on tumor induction and RNA synthesis in *Agrobacterium tumefaciens.* Microbios *23:* 115–126 (1978).
Sollner-Webb, B.; Felsenfeld, G.: A comparison of the digestion of nucleic chromatin by staphylococcal nuclease. Biochemistry *14:* 2915–2920 (1975).
Soprano, K.J.; Baserga, R.: Reactivation of ribosomal RNA genes in human hybrid cells by 12-O-tetradecanoylphorbol-13-acetate. Proc. natn. Acad. Sci. USA *77:* 1566–1569 (1980).
Sperling, L.; Tardieu, A.; Weiss, M.C.: Chromatin repeat length in somatic hybrids. Proc. natn. Acad. Sci. USA *77:* 2716–2720 (1980).
Spirin, A.S.: Informosomes. Eur. J. Biochem. *10:* 20–35 (1969).
Srivastava, B.I.S.: Inhibition of oncorna-virus and cellular DNA polymerases by natural and synthetic polynucleotides. Biochim. biophys. Acta *335:* 77–84 (1973).
Stark, G.R.; Dower, W.J.; Schimke, R.T.; Brown, R.E.; Kerr, I.M.: 2-5 synthetase: assay, distribution and variation with growth or hormone status. Nature, Lond. *278:* 471–473 (1979).
Stead, N.W.; Jones, O.W.: The binding of RNA polymerase to DNA: stabilization by nucleoside triphosphates. Biochim. biophys Acta *145:* 679–688 (1967).
Stedman, E.; Stedman, E.: Cell specificity of histones. Nature, Lond. *166:* 780–781 (1950).
Stockdale, F.E.; Topper, Y.J.: The role of DNA synthesis and mitosis in hormone dependent differentiation. Proc. natn. Acad. Sci. USA *56:* 1283–1289 (1966).
Strauss, J.H., Jr.; Kelly, R.B.; Sinsheimer, R.L.: Denaturation of RNA with dimethylsulfoxide. Biopolymers *6:* 793–807 (1968).
Strauss, N.A.; Birnboin, H.C.: Polypyrimidine sequences found in eukaryotic DNA have been conserved during evolution. Biochim. biophys. Acta *454:* 419–428 (1976).
Stravianopoulos, J.G.; Karkas, J.D.; Chargaff, E.: Nucleic acid polymerases of the developing chicken embryos: a DNA polymerase preferring a hybride template. Proc. natn. Acad. Sci. USA *68:* 2207–2211 (1971).
Stravianopoulos, J.G.; Karkas, J.D.; Chargaff, E.: Mechanism of DNA replication by highly purified DNA polymerase of chicken embryos. Proc. natn. Acad. Sci. USA *69:* 2609–2613 (1972).
Stroun, M.; Anker, P.; Gahan, P.; Rossier, A.; Greppin, H.: *Agrobacterium tumefaciens* ribonucleic acid synthesis in tomato cells and crown-gall induction. J. Bact. *106:* 634–639 (1971).
Strumpf, W.E.: Estradiol-concentrating neurons: topography in the hypothalamus by dry mount autoradiography. Science *162:* 1001–1003 (1968).

Sturgess, E.A.; Ballantine, J.E.M.; Woodland, H.R.; Wohun, P.R.; Lane, C.D.; Dimitriados, G.J.: Actin synthesis during early development of *Xenopus laevis.* J. Embryol. exp. Morph. *58:* 303–320 (1980).

Sugimoto, M.; Tajima, K.; Kojima, A.; Endo, H.: Differential acceleration by hydrocortisone of the accumulation of epidermal structural proteins in the chick embryonic skin growing in chemically defined medium. Devl Biol. *39:* 295–307 (1974).

Svoboda, D.; Reddy, J.; Harris, C.: Invasive tumors induced in rats with actinomycin D. Cancer Res. *30:* 2271–2279 (1970).

Takeshita, M.; Grollman, A.P.; Ohtsubo, E.; Ohtsubo, H.: Interaction of bleomycin with DNA. Proc. natn. Acad. Sci. USA *75:* 5983–5987 (1978).

Tanaka, M.; Levy, J.; Terada, M.; Breslow, R.; Rifkind, R.A.; Marks, P.A.: Induction of erythroid differentiation in murine virus infected erythroleukemic cells by highly polar compounds. Proc. natn. Acad. Sci. USA *72:* 1003–1006 (1975).

Tanaka, M.; Yoshida, S.: Mechanism of the inhibition of calf thymus DNA polymerases α and β by daunomycin and adriamycin. J. Biochem. *87:* 911–918 (1980).

Tashima, M.; Sawada, H.; Tatsumi, K.; Nakamura, T.; Uchino, H.: Effect of 1,3-bis(2-chloroethyl)-*l*-nitrosourea on newly synthesized DNA in L 1210 cells. Biochem. Pharmac. *28:* 511–517 (1979).

Teissere, M.; Penon, P.; Ricard, J.: Hormonal control of chromatin availability and of the activity of purified RNA polymerase in higher plants. FEBS Lett. *30:* 65–70 (1973).

Temin, H.M.: Mechanism of cell transformation by RNA tumor viruses. A. Rev. Microbiol. *25:* 609–648 (1971).

Temin, H.M.; Mizutani, S.: RNA dependent DNA polymerases in virions of Rous sarcoma virus. Nature, Lond. *226:* 1211–1213 (1970).

Teng, C.S.; Hamilton, T.H.: The role of chromatin in estrogen action in the uterus. I. The control of template capacity and chemical composition and the binding of 17-β H^3 estradiol. Proc. natn. Acad. Sci. USA *60:* 1410–1417 (1968).

Thompson, E.B.; Granner, D.K.; Tomkins, G.M.: Superinduction of tyrosine aminotransferase by actinomycin D in rat hepatoma cells (HTC). J. molec. Biol. *54:* 159–175 (1970).

Thrash, C.R.; Cunningham, D.D.: Stimulation of division of density inhibited fibroblasts by glucocorticoids. Nature, Lond. *242:* 399–401 (1973).

Thurnherr, N.; Deschner, E.E.; Stonehill, E.H.; Lipkin, M.: Induction of neurocarcinomas of the colon in mice by weekly injections of 1,2-dimethylhydrazine. Cancer Res. *33:* 940–945 (1973).

Tice, L.W.; Barrnett, R.J.: The fine structural localization of glucose-6-phosphatase in rat liver. J. Histochem. Cytochem. *10:* 754–762 (1962).

Timmis, J.V.; Ingle, J.: Environmentally induced changes in rRNA gene redundancy. Nature new Biol. *244:* 235–236 (1973).

Tomkins, G.M.; Levinson, B.B.; Baxter, J.D.; Dethlefsen, L.L.: Further evidence for posttranscriptional control of inducible tyrosine amino-transferase synthesis in cultured hepatoma cells. Nature new Biol. *239:* 9–14 (1972).

Trachewsky, D.; Segal, S.J.: Differential synthesis of ribonucleic acid in uterine nuclei; evidence for selective gene transcription induced by estrogens. Eur. J. Biochem. *4:* 279–285 (1968).

Trewavas, A.J.: Effect of IAA on RNA and protein synthesis. Archs. Biochem. Biophys. *123:* 324–335 (1968).

Troll, W.; Klassen, A.; Janoff, A.: Tumorigenesis in mouse skin: inhibition by synthetic inhibitors of proteases. Science *169:* 1211–1213 (1970).

Tsai, S.T.; Tsai, M.J.; Lin, C.T.; O'Malley, B.W.: Effect of estrogen on ovalbumin gene expression in differentiated nontarget tissue. Biochemistry *18:* 5726–5731 (1979).

Tseng, B.Y.; Goulian, M.: Initiator RNA of discontinuous DNA synthesis in human lymphocytes. Cell *12:* 483–489 (1977).

Ts'o, P.O.P.; Helmkamp, G.K.; Sander, C.: Interaction of nucleosides and related compounds with nucleic acids as indicated by the change of helix-coil transition temperature. Proc. natn. Acad. Sci. USA *48:* 686–698 (1962a).

Ts'o, P.O.P.; Helmkamp, G.K.; Sander, C.: Secondary structure of nucleic acids in organic solvents. II. Optical properties of nucleotides and nucleic acids. Biophys. biochim. Acta *55:* 584–600 (1962b).

Tung, T.C.; Niu, M.C.: Nucleic acid induced transformation of goldfish. Scientia sinica *15:* 377–384 (1973).

Tung, T.C.; Niu, M.C.: Transmission of the nucleic acid-induced character, caudal fin, to the offspring in goldfish. Scientia sinica *18:* 223–228 (1975).

Tuonunen, F.W.; Kenney, F.T.: Inhibition of the DNA polymerase of Rauscher leukemia virus by single-stranded polyribonucleotides. Proc. natn. Acad. Sci. USA *68:* 2198–2202 (1971).

Umezawa, H.: Bleomycin: discovery, chemistry and action. Gann. Monogr. Cancer Res. *19:* 3–36 (1976).

Van de Sande, J.H.; Lin, C.C.; Jorgenson, K.F.: Reverse banding on chromosomes produced by a guanosine-cytosine specific DNA binding antibiotic: olivomycin. Science *195:* 400–402 (1977).

Vatanova, H.; Blaha, K.; Sponar, J.: Model of nucleoproteins: binding of actinomycin D to complex of DNA with lysine containing polypeptides. Biopolymers *17:* 1747–1758 (1978).

Villee, C.A.; Goswani, A.: Effect of exogenous RNA on steroid metabolism in adrenals and gonads; in Niu, Segal, The role of RNA in reproduction and development, pp. 75–85 (North Holland, Amsterdam 1973).

Waddington, C.H.: New patterns in genetics and development (Columbia University Press, New York 1962).

Wagar, M.A.; Huberman, J.A.: Covalent linkage between RNA and nascent DNA in mammalian cells. Cell *6:* 551–557 (1975).

Walter, G.; Zillig, W.; Palm, P.; Fuchs, E.: Initiation of DNA-dependent RNA synthesis and the effect of heparin on RNA polymerase. Eur. J. Biochem. *3:* 194–201 (1967).

Walters, R.A.; Tobey, R.A.; Hildebrand, C.E.: Hydroxyurea does not prevent synchronized G_1 chinese hamster cells from entering the DNA synthetic period. Biochim. biophys. Res. Commun. *69:* 212–217 (1976).

Walther, B.J.; Pictet, R.L.; David, J.D.; Rutter, W.J.: On the mechanism of 5-bromodeoxy-uridine inhibition of exocrine pancreas differentiation. J. biol. Chem. *249:* 1953–1964 (1974).

Wang, A.H.J.; Quigley, G.J.; Kolpak, F.J.; Crawford, J.L.; Boom, J.H. von; Van der Marel, G.; Rich, A.: Molecular structure of a left-handed double helical DNA fragment at atomic resolution. Nature, Lond. *282:* 680–686 (1979).

Wang, J.C.: Unwinding of DNA by actinomycin binding. Biochim. biophys. Acta *232:* 246–251 (1971).

Wang, J.C.; Jacobsen, J.H.; Saucier, J.M.: Physicochemical studies on interactions between DNA and RNA polymerase. Unwinding of the DNA helix by *Escherichia coli* RNA polymerase. Nucl. Acids Res. *4:* 1225–1241 (1977).

Waring, M.J.: Complex formation between ethidium bromide and nucleic acids. J. molec. Biol. *89:* 783–801 (1965).

Waring, M. J.: Variation of the supercoils in closed circular DNA by binding of antibiotics and drugs. Evidence for molecular models involving interaction. J. molec. Biol. *54:* 257–279 (1970).

Warwick, G.P.: The covalent binding of metabolites of tritiated 2-methyl-4-dimethylaminobenzene to rat liver nucleic acids and proteins and the carcinogenicity of the unlabelled compound in partially hepatectomized rats. Eur. J. Cancer *3:* 227–233 (1967).

Wasylyk, B.; Chambon, P.: Studies on the mechanism of transcription of nucleosomal complexes. Eur. J. Biochem. *103:* 219–226 (1980).

Watanabe, T.; Taguchi, Y.; Sasaki, K.; Tsuyama, K.; Kitamura, Y.: Increase in histidine decarboxylase activity in mouse skin after application of the tumor promoter tetra-decanoylphorbolacetate. Biochem. biophys. Res. Commun. *100:* 427–432 (1981).

Watson, J.D.: Molecular biology of the gene (Benjamin, New York 1965).

Watson, J.D.; Crick, F.H.C.: Molecular structure of nucleic acids. Nature, Lond. *171:* 735–738 (1953).

Watters, C.; Gullino, P.M.: Translocation of DNA from the vascular into the nuclear compartment of solid mammary tumors. Cancer Res. *31:* 1231–1243 (1971).

Weckler, C.; Gschwendt, M.: The effect of estradiol on the activity of the nucleolar and nucleoplasmic RNA polymerases from chicken liver. FEBS Lett. *65:* 220–224 (1976).

Weil, P.A.; Sidikaro, J.; Stancel, G.M.; Blatti, S.P.: Hormone control of transcription in the rat uterus. J. biol. Chem. *252:* 1092–1098 (1977).

Weinberg, R.A.: Nuclear RNA metabolism. A. Rev. Biochem. *42:* 329–354 (1973).

Weiner, J.H.; McMacken, R.; Kornberg, A.: Isolation of an intermediate which precedes DNA G RNA polymerase participation in enzymatic replication of bacteriophage *Φ*X174 DNA. J. biol. Chem. *250:* 1972–1980 (1975).

Weinmann, R.; Roeder, R.G.: Role of DNA-dependent RNA polymerase III in transcription of the t-RNA and 5S RNA genes. Proc. natn. Acad. Sci. USA *71:* 1790–1799 (1974).

Weinstein, I.B.; Wigler, M.; Fisher, P.B.; Siskin, E.; Pietropolo, C.: Cell culture studies on the biologic effect of tumor promoters; in Slaga, Sivak, Boutwell, Mechanism of tumor promotion and cocarcinogenesis: a comprehensive survey, vol. 2, pp. 313–333 (Raven Press, New York 1978).

Weinstock, R.; Sweet, R.; Weiss, M.; Cedar, H.; Axel, R.: Intragenic DNA spacers interrupt the ovalbumin gene. Proc. natn. Acad. Sci. USA *75:* 1299–1303 (1978).

Weintraub, H.; Worcel, A.; Alberts, B.: A model for chromatin based upon symmetrically paired half-nucleosomes. Cell *9:* 409–417 (1976).

Weir, D.R.; Malher, V.L.: The occurrence of apparent chronic myeloid leukemia in mice following intermittent cortisone therapy. J. Lab. clin. Med. *42:* 963 (1953).

Weissbach, A.; Baltimore, D.; Bollum, F.; Gallo, R.; Corn, T.: Nomenclature of eukaryotic DNA polymerases. Eur. J. Biochem. *59:* 1–2 (1975).

Weissbach, A.; Schlabach, A.; Fridlender, B.; Bollen, A.: DNA polymerases from human cells. Nature new Biol. *231:* 167–170 (1971).

Wells, R.D.; Flügel, R.M.; Larson, J.E.; Schendel, P.F.; Sweet, R.W.: Comparison of some reactions catalysed by deoxyribonucleic acid polymerase from avian myeloblastosis virus, *Escherichia coli* and *Micrococcus luteus.* Biochemistry *11:* 621–629 (1972).

Wessells, N.K.; Wilt, F.H.: Action of actinomycin D on exocrine pancreas cell differentiation. J. molec. Biol. *13:* 767–779 (1965).

Whitlock, J.P., Jr.; Simpson, R.T.: Removal of histone H_1 exposes a fifty base pair DNA segment between nucleosomes. Biochemistry *15:* 3307–3313 (1976).

Wigler, M.; Weinstein, I.B.: Tumor promoter induces plasminogen activation. Nature, Lond. *259:* 232–259 (1976).

Williams-Ashman, H.G.; Jurkowitz, L.; Silverman, D.A.: Testicular hormones and the synthesis of ribonucleic acids and proteins in the prostate gland. Recent Prog. Horm. Res. *20:* 247–301 (1964).

Williams, P.H.; Bayer, H.W.; Helinski, D.K.: Size and base composition of RNA in supercoiled plasmid DNA. Proc. natn. Acad. Sci. USA *74:* 3744–3748 (1973).

Wilson, C.B.; Boldrey, E.B.; Enot, K.J.: 1,3-Bis(2-chloroethyl)-*l*-nitrosourea (NSC-409962) in the treatment of brain tumors. Cancer Chemother. Rep. *54:* 273–281 (1970).

Witkin, E.M.: DNA repair and mutagenesis; in Mécanismes d'altération et de reparation du DNA. Relation avec la cancérogénèse chimique, pp. 203–225 (CNRS, Paris 1977).

Witkin, S.S.; Korngold, G.C.; Bendich, A.: Ribonuclease sensitive DNA-synthesizing complex in human sperm heads and seminal fluid. Proc. natn. Acad. Sci. USA *72:* 3295–3299 (1975).

Wittliff, J.L.; Lee, K.L.; Kenney, F.T.: Regulation of yolk protein synthesis in amphibian liver. II. Elevation of ribonucleic acid synthesis by estrogen. Biochim. biophys Acta *269:* 493–504 (1972).

Woese, C.R.: The genetic code; the molecular basis for genetic expression (Harper & Row, New York 1967).

Wolsky, A.: The effect of chemicals with gene-inhibiting activity on regeneration; in Sherbet, Neoplasia and cell differentiation, pp. 153–188 (Karger, Basel 1974).

Wolsky, A.: Regeneration and cancer (Editorial). Growth *42:* 425–426 (1978).

Wolsky, A.; de Issekutz-Wolsky, M.: Molecular-biological aspects of epigenetics and their phylogenetic implications; in Novak, Mlikovsky, Evolution and environment. Proceedings of the International Symposium Brno. Czechoslovak Academy of Sciences, Praha, pp. 51–58 (1982).

Wreschner, D.H.; McCauley, J.W.; Skehel, J.J.; Kerr, I.M.: Interferon action sequence specificity of the ppp(A2′p)nA-dependent ribonuclease. Nature, Lond. *289:* 414–417 (1981).

Wu, A.M.; Gallo, R.C.: Reverse transcriptase. CRS crit. Rev. Biochem. *1975:* 284–347 (1975).

Wu, C.W.; Goldthwait, D.A.: Studies of nucleotide binding to the ribonucleic acid polymerase by a fluorescence technique . Biochemistry *8:* 4450–4457 (1969a).

Wu, C.W.; Goldthwait, D.A.: Studies of nucleotide binding to the ribonucleic acid polymerase by equilibrium dialysis. Biochemistry *8:* 4458–4464 (1969b).

Yamamoto, K.R.: Characterization of the 4S and 5S forms of the estradiol receptor protein and their interaction with deoxyribonucleic acid. J. biol. Chem. *249:* 7068–7075 (1974).

Yarrouton, G.T.; Das, R.H.; Gefter, M.L.: Enzyme-catalysed DNA unwinding. Mechanism of action of helicase III. J. biol. Chem. *254:* 12002–12006 (1979).

Yu, F.L.: High concentration of RNA polymerase I is responsible for the high rate of nucleolar transcription. Biochem. J. *188:* 381–385 (1980).

Yu, F.L.; Feigelson, P.: The sequential stimulation of uracil-rich and guanine-rich RNA species during cortisone induction of hepatic enzymes. Biochem. biophys. Res. Commun. *35:* 499–504 (1969).

Yuspa, S.H.; Ben, T.; Hennings, H.; Lichti, U.: Phorbol ester tumor promoters induce epidermal transglutaminase activity. Biochem. biophys. Res. Commun. *97:* 700–708 (1980).

Zeichman, M.; Brutkrentz, D.: Isolation of low molecular weight RNAs from connective tissue. Inhibition of mRNA translation. Archs Biochem. Biophys. *188:* 410–417 (1978).

Zetterberg, A.; Engström, W.: Mitogenic effect of alkaline pH on quiescent, serum starved cells. Proc. natn. Acad. Sci. USA *78:* 4334–4338 (1981).

Zieve, G.; Penman, S.: Small RNA species on the HeLa cell metabolism and subcellular localization. Cell *8:* 19–31 (1976).

Zimmer, C.H.: Netropsin and distamycin A. Prog. nucleic Acid Res. mol. Biol. *15:* 285–318 (1975).

Zimmer, C.H.; Preschendorf, B.; Grunicke, H.; Chandra, P.; Venner, H.: Influence of Netropsin and distamycin A on the secondary structure and template activity of DNA. Eur. J. Biochem. *21:* 269–278 (1971).

Zylber, E.A.; Penman, S.: Products of RNA polymerases in HeLa cell nuclei. Proc. natn. Acad. Sci. USA *68:* 2861–2865 (1971).

Subject Index